GenEthics and Religion

GenEthics and Religion

Editors

G. Pfleiderer Basel
G. Brahier Basel
K. Lindpaintner Newark, Del.

KARGER

Basel · Freiburg · Paris · London · New York · Bangalore · Bangkok · Shanghai · Singapore · Tokyo · Sydney

G. Pfleiderer
Department of Theology
University of Basel
Missionsstrasse 17a
CH–4056 Basel
Switzerland

G. Brahier
Department of Theology
University of Basel
Missionsstrasse 17a
CH–4056 Basel
Switzerland

K. Lindpaintner
SDIX
111 Pencader Drive
Newark, DE 19702
USA

Library of Congress Cataloging-in-Publication Data

GenEthics and religion / volume editors, G. Pfleiderer, G. Brahier, K. Lindpaintner.
p. ; cm.
Includes bibliographical references and indexes.
ISBN 978-3-8055-8973-4 (hard cover : alk. paper)
1. Genetic engineering--Religious aspects--Congresses. I. Pfleiderer, Georg. II. Brahier, G. (Gabriela) III. Lindpaintner, K. (Klaus)
[DNLM: 1. Genetics--ethics--Congresses. 2. Bioethical Issues--Congresses. 3. Religion and Medicine--Congresses. QU 450 G3272 2010]
QH438.7.G4116 2010
205'.64957--dc22

2010014350

Bibliographic Indices. This publication is listed in bibliographic services, including Current Contents® and Index Medicus.

www.karger.com
Printed in Switzerland on acid-free and non-aging paper (ISO 9706) by Reinhardt Druck, Basel
ISBN 978–3–8055–8973–4
e-ISBN 978–3–8055–8974–1

Contents

Introduction

1 **Beyond Playing God: Critical Religious GenEthics for Pluralistic Societies**
Pfleiderer, G.; Brahier, G. (Basel); Lindpaintner, K. (Newark, Del.)

Foundations: Hermeneutic and Conceptional Reflections

12 **Genes – Cells – Interpretations: What Hermeneutics Can Add to Genetics and to Bioethics**
Rehmann-Sutter, C. (Lübeck)

28 **Controversies about Human Dignity: Implications for Biotechnology**
Childress, J.F. (Charlottesville, Va.)

The Function of Religion in GenEthical Debates: Critical Analyses

40 **Global Bioethics, Theology, and Human Genetic Engineering: The Challenge of Refashioning Human Nature in the Face of Moral and Religious Pluralism**
Engelhardt, Jr., H.T. (Houston, Tex.)

52 **Eschewing Images of Man: Against Anthropological Reductionism in Bioethics**
Graf, F.W. (Munich)

60 **Children, Bodies, Life: Ethics as the Churches' Biopolitics**
Gehring, P. (Darmstadt)

74 **On the Ethics Debate between Theologians, Scientists and Doctors: Experiences, Observations and Commentaries of a Medical Geneticist**
Müller, H. (Basel)

Examining Constructive Efforts of Religious GenEthics (I): Christian Theological GenEthics

82 **'Biopower' – Ethical and Theological Considerations**
Mieth, D. (Erfurt)

101 **First Sheep, Then Human Beings? Theological and Ethical Reflections on the Use of Gene Technology**
Schockenhoff, E. (Freiburg)

108 **Is the Human Genome Sacred?**
Peters, T. (Berkeley, Calif.)

Examining Constructive Efforts of Religious GenEthics (II): Jewish, Islamic, and Buddhist GenEthics

118 **The Jewish Perspective on GenEthics**
Green, R.M. (Hanover, N.H.)

129 **Human Genetic Technologies and Islamic Bioethics**
Nor, S.N.M. (Kuala Lumpur)

138 **Buddhism and Human Genome Research**
Ratanakul, P. (Nakornpathom)

151 **Author Index**

152 **Subject Index**

Pfleiderer G, Brahier G, Lindpaintner K (eds): GenEthics and Religion. Basel, Karger, 2010, pp 1 11

Beyond Playing God: Critical Religious GenEthics for Pluralistic Societies

Georg Pfleiderer[a] · Gabriela Brahier[a] · Klaus Lindpaintner[b]

[a]Department of Theology, University of Basel, Basel, Switzerland; [b]SDIX, Newark, Del., USA

Playing God?

Human gene technology or human gene engineering is a diverse and rapidly developing sphere of research and advanced medical technology. It includes stem cell research, genetic testing, therapeutic or reproductive cloning. Since genes are something like the blueprints of organisms, genetic engineering is a challenge to our understanding of human nature. Thus, from the early days of genetic research, projects have been watched carefully and often suspiciously by ethicists. Among this new caste of bioethicists dealing with these problems theologians have played a prominent role. Not surprisingly: as fundamental questions are raised with regard to whether human beings should be allowed to manipulate their own species, a religious dimension of human actions seems to be touched. Is genetic engineering by definition playing God? And if so, why is it fundamentally different from other areas where human ingenuity profoundly affects our lives?

The volume examines questions pertaining to the complex relationship between the ethics of gene technology (moniker: *genethics*) and religion. It does so in a critical and multidisciplinary way, and with reference to several of the major world religions. The volume contains contributions by philosophers, by scholars of divinity, and by bioethicists with Christian, Jewish, Islamic, and Buddhist background.

In this introduction, use has been made of unpublished abstracts which were submitted for the conference by the following authors: J. Childress, H.T. Engelhardt Jr., S.N.M. Nor, P. Ratanakul, and C. Rehmann-Sutter. Note that this text, in part, contains literal quotes from these abstracts as well as from the authors' articles contained in this volume; for ease of reading, these quotes have not been set in quotation marks.

Foundations: Hermeneutic and Conceptional Reflections

Ethical evaluations should start with hermeneutical reflections, particularly if the genome is the object of interpretation. This is at least the suggestion of **Christoph Rehmann-Sutter**. The operative function of the genome for the organism is both a biological and a descriptive problem. The respective terminology, used not only by journalists but also by scientists, is full of metaphors. Textual metaphors are most common: genes are described as *texts*, *instructions*, *messages*, *blueprints* or *programs*. The textual imagery, however, seems to be misleading: the genome is not an author-subject, neither is it – as at least some of these metaphors suggest – strictly prescriptive or deterministic. Thus, Rehmann-Sutter pleads for a shift towards a new metaphor: *systems*. A systems approach to genomics opens the way for new images: DNA as a big library that can be used in many different ways and whose content depends on the context of use; or the multifunctionality of genes being compared to words that have different meanings. He argues that the transition from program genomics to systems genomics has implications for central topics of bioethics like genetic testing and counseling, or the moral perception of human embryos.

Critical reflection must also be used regarding the normative concepts of genethics. One of the most frequently used normative concepts is human dignity. The concept suggests an absolute, intrinsic, and ontological value of human beings and their nature. Opponents argue that this is a classical proof of naturalistic fallacy or of an overabundance of normative rhetoric. In order to avoid both, **James Childress** proposes a clear analysis of the concept. He distinguishes between (1) human dignity as a standard of rights and duties, and (2) human dignity as a standard of virtue. The first concept often goes along with the Kantian prohibition of instrumentalization. While the first would refer more to the person as the object of a gene-technological manipulation or treatment, the second is more focused on the character – the virtues or vices – of the users or providers of human engineering.

Although Childress himself favors a reflected and cautious use of the first understanding of the concept, he highlights at least to two problems: what is the added value of the concept compared, e.g., with more rational concepts such as autonomy? And how does the concept work as a normative ethical instrument in answering concrete bioethical questions? For instance: it is hard to make the case that human reproductive cloning necessarily involves the kind of instrumentalization that human dignity excludes. Human reproductive cloning does not necessarily involve treating offspring merely as means or instruments to other ends. The argument shows that it seems to be difficult to concentrate on the perspective of the object-person.

Concerning the virtue-concept Childress refers strongly to Leon Kass and the (US) President's Council on Bioethics (during the Bush Administration, suspended in 2009 by President Obama). In Kass' concept human dignity and its violation is based on a fundamentally human pre-reflective moral wisdom of repugnance. Examples of wise repugnance include bestiality, cannibalism, mutilation of a corpse, and raping or murdering another human being. Repugnance is sufficient to condemn these actions; without rational justification, the repugnance at human cloning belongs in this category. In this understanding human dignity seems to work as a fundamental moral taboo, referring to a religious dimension of morality. This concept of human dignity is therefore suspected to smuggle in religious and theological concepts without having to acknowledge them as such. As similar observations could be made in the field of ethical debates on human genetic enhancement, Childress pleads for a public precautionary process in which society's basic values and the public's reflective preferences are carefully debated and idiomatic manipulations minimized.

The Function of Religion in GenEthical Debates: Critical Analyses

H. Tristram Engelhardt and Friedrich Wilhelm Graf agree with James Childress that the common tendencies to involve religious arguments in genethical debates have to do with the pressing character of the moral problems at hand. However, they appear to be less confident regarding the potential chances of success of the public precautionary process that Childress advocates.

For **H. Tristram Engelhardt** the possibility of human genetic engineering functions like a Rorschach test that discloses foundational disagreements regarding the nature of morality, human flourishing, and the significance of human existence. It discloses our deep moral disagreements. The human condition is characterized by culture wars, by foundational disputes about how humans should live, and about how they should apply technologies such as human genetic engineering. Thus, in his chapter Engelhardt argues that there is no common, secular morality that can rationally establish a single canonical, content-rich, global code of bioethics to guide our emerging human gene-technology capacities. Second, he contends that there is no way to set aside these secular moral disagreements through rational argument. Third, he attributes the fact that there are still claims of moral consensus mostly to wishful thinking or to political and other interests.

Engelhardt seems to be convinced that the impossibility of a true moral consensus on secular grounds is the logical consequence of the absence of an ultimate metaphysical principle of life and of our existence. However, he is aware that in the realm of religion moral pluralism, particularly with regard to human genetic engineering, is all but overcome. On the contrary, mutual religious anathematism rather intensifies than appeases moral dissent in genethics.

Nevertheless, these findings are viewed as encouraging rather than preventing moral statements based on outspoken confessional-religious background. A self-professed member of the Christian Orthodox church, Engelhardt therefore sketches Orthodoxy's understandings of the norms that should guide human genetic engineering so as to offer a more complete view of the moral diversity that characterizes our condition, as well as to present a theologically grounded approach to human genetic engineering that is, in the main, very positive. The reason for Christian Orthodoxy's surprising openness towards modern gene technology is, according to Engelhardt, the fact that orthodox religious ethic is oriented towards Eden before the Fall; therefore human enhancement is considered to be of vital necessity for all human beings in the present stage, i.e., after the fall.

The balls of impregnable religious pluralism and of the putative vanity of an objective judgment of religious truth claims are picked up by **Friedrich Wilhelm Graf**. Notwithstanding this skeptical attitude, like Engelhardt he does not argue for a general abandonment of theological bioethics, but for an open declaration of individual religious standpoints and traditions, in his case founded in a liberal Protestant culture of Lutheran tradition. Despite his radical noological and dogmatic skepticism he believes in the enlightenment and consensual power of historical critique. Thus, he conceives of a historical critique of religious semantics and concepts in moral debates on topics such as genetic engineering, as a major task of academic bioethics.

Current public bioethical debates, he contends, would be intensively contaminated by the application of religious overtones. In Graf's eyes an involvement of religious semantics is highly counterproductive to ethical debates; as it is a method of 'suggesting authority' which would be problematic as he contends that religious ideas are probably the most dangerous mental substance that we humans constantly have to deal with. Religious language is extremely flexible and prone to abuse. The metaphor of creation is very popular in debates on genetics, and it is used for discrediting any kind of genetic engineering, as well as for the exact contrary.

According to Graf, it is also typical for religious language to transcend its original context and infiltrate public secular moral language. Using the tools of a critical history of concepts, populist claims of an absolute human dignity or a Christian or Western conception of man *(Menschenbild)* with normative implications as well as references made to the world as a whole may represent strategies of political-religious semantics which would be of little help in resolving current bio-political problems. If there is a sustainable claim for religious truth with respect to anthropology and bioethics, then it is the idea of the fragmentary character of life and the tenet that man is to be defined as indefinable.

Petra Gehring is unlikely to have Friedrich Wilhelm Graf in mind when she points to a new order *(caste)* of professional theological bioethicists, raised over the last 30 years by the churches (in cooperation with academic theology) as effective instruments to assert themselves as social agents of bio-power. Like Graf she tries to support her general thesis using a historical analysis of discourses, in her case of churches' statements and activities in discussions on abortion, xenotransplantation, brain death, and advance directives/living wills.

Although her contribution does not focus exclusively on the bioethical aspects of genetics, she holds as a central argument that the churches were strong forces in transforming the respective public health debates more and more into discourses on life and biology itself. In this sense *genetization* would be the extreme of the general *bioethicization* of the churches' positions.

This, she argues, happened first in the discourse on abortion when the Roman Catholic Church in particular used its interpretation of a biological definition of life (which had already been coined in 19th century) to establish a rather patronizing position in the sense of a virtually absolute protection of life that ignored the woman almost entirely, and aimed at protecting the embryo as an entity separate from the mother. This theological development seems to correspond to major trends in recent public debates on moral issues. It represents a shift from earlier 20th century theological debate by both Catholic and Protestant denominations when arguments were based less on an understanding of the nature of life, and more on social relationships (family, national ties).

According to Gehring, this new theological tendency towards a science-based understanding of life is also responsible for the unexpectedly positive stance which Christian churches took towards xenotransplantation, and with regard to the closely related debate on brain death as well as the more recent debates on advance directives or living wills. The fact that the initial reaction of the German churches in this respect was characterized by a remarkable indifference is, in Gehring's eyes, additional evidence that the evolving ecclesiastical and theological view of human ethics is not based on a new theological metaphysic, but rather on the functionality of biology-grounded arguments and general adaption strategies.

The geneticist **Hansjakob Mueller** describes the discourse between theologians and natural scientists on ethical questions of modern genetics from his personal lifelong experience, rather than from a historical-analytical perspective. He is convinced that both disciplines share a mutual interest in the wonders and preservation of human nature, and that there is clearly the need for interdisciplinary ethical dialogue. Nevertheless, Mueller is skeptical about the role that theology can play in this regard. In his eyes theologians should be more open to the progress of scientific research without fear of weakening theological moral principles. On the other hand, he is also wary of a quasi-professionalization of theological bioethicists. Thus Mueller sees the task of theological ethics, as well as professional ethics in general, not in the reflection of general principles but in the formulation of clear guidelines for complex but concrete ethical issues. Mueller concludes that the diversity of approaches, both within natural sciences and within theology, must be taken seriously and be integrated in a constructive fashion into the interdisciplinary ethics debate.

Examining Constructive Efforts of Religious Genethics

Christian Theological Genethics

Critical analyses of theological or the churches' bio-power are – as Friedrich Wilhelm Graf already convincingly argued – not the privilege of non-theological observers.

In his chapter the Roman-Catholic theologian **Dietmar Mieth**, a strong representative of advanced theological bioethics, opens the focus of critical observation towards a more general analysis of bio-political instruments and semantics. In his eyes bio-power is not a quasi-subjective entity in the sense of Michel Foucault, but rather the interaction of various agents and stakeholders in the field of bioethics. For democratic society, he holds, the balancing powers and interests in a field of such vital interest must be as open as possible to public dialogue.

According to Mieth, prototypical instruments of bio-political maneuvering are moratoria. Moratoria may allow room for better ethical consideration and new empirical facts, but they are also never innocent in the sense that they usually are a first step on the path toward an ethical compromise. A good example for the ambiguous character of this political tool was the discussion on human cloning held in the Council of the European Union, particularly in its Advisory Group on Ethical Implications of Biotechnology in the 1990s. Regarding this subject, Mieth argues for a mediated position based on a selection of arguments. Although each of these arguments would be relative, Mieth is convinced that in sum they force not only a moratorium, but a ban on human cloning.

A second typical element and instrument of bio-power are the politics of language. A good example, also from the field of human cloning, is the differentiation between in vivo and in vitro cloning in combination with the terms reproductive and non-reproductive or even therapeutic cloning. Although such differentiation is necessary, Mieth is convinced that in many debates it tends to complicate rather than solve problems – he cites as an example the Human Rights Convention for Biomedicine. Still, Mieth argues for a prohibition of reproductive and in vivo cloning, but for greater leniency regarding other forms.

Mieth argues that analyzing the elements of bio-power does not interfere with taking a position within the respective debates, lest theology in particular would lose its very character. Here again Mieth seeks to combine an emphatic and principalist position (a practical Kantian apriorism) that – in his eyes – demands absolute protection of the embryo and its dignity with a combination of arguments. Concerning the question of human cloning, such arguments include the idea that humans are the image of God, as well as the diversity of humanity, finiteness, and the vulnerability of humanity. Each of these theological arguments, he concedes, is relative, but in their sum they are compelling. Applied to the question of the ethical status of the early embryo, he holds that this argues clearly for the official position of the Roman Catholic magisterium, i.e., the assumption that an embryo is to be treated like a person because of the unity and the continuity of human development.

Bio-power is not only a controversial issue, but also one of real political forces. Therefore, Mieth concludes his chapter with a hint at the dangers of an instrumentalization of women by invasive medical techniques of fertilization and at the powerful economic interests associated with these techniques. Hence, he argues for a strategic alliance of moral theology with a new form of feminism.

A bio- and genethical position which is quite close to that of Dietmar Mieth is presented by another Roman Catholic moral theologian and bioethicist, **Eberhard Schockenhoff**. He is convinced that – even on theological grounds – there is no principal verdict against gene technology in general. It is evident for him, however, that at least the most extreme interventions that gene technology potentially provides, i.e. the cloning of an entire human individual, are associated with the most serious ethical problems. However, less radical applications of the technology, such as genetic diagnostics are also not without ethical problems or a priori admissible.

Schockenhoff's foundational normative conviction in this context is represented by the Kantian idea that human dignity is based on the principal and unconditional acceptance of the individual per se. Schockenhoff is also a determined defender of a continuity argument with regard to human development from its earliest stages. Nevertheless, in his contribution to this volume he reserves the principal deontological arguments mostly for issues surrounding genetic manipulation. Regarding prenatal diagnosis, his arguments are mostly consequentialist. Thus, he refers to practical problems such as the risk of damaging an embryo, and to possible socio-psychological effects such as the predicted decline of social acceptance of children with congenital disabilities and the growing pressure on parents to prevent giving birth to handicapped children. Thus, he pleads for, at the very least, a *right not to know*.

He holds that the cloning of human beings is incompatible with the notion of the personal dignity and individual purpose of each human being. The systematic argument carrying this normative statement rests on the idea that human freedom is a finite freedom, and that it is based on a natural condition which must, in its genetic basis, not be modulated by intentional acts of other individuals. Any intentional interference in the genetic structure would – as Schockenhoff agrees with Jürgen Habermas[1] – unduly limit the chance for an unimpeded development and self-understanding of the individual.

Confronted with such rather restrictive genethical reasoning, more liberal Protestant theologians like **Ted Peters** ask the question (as he does in the title of his essay): Is the Human Genome Sacred? There are tendencies in public opinion and as well in ethical and theological debates to treat the genome as the essential substrate of human nature and thereby as a quasi-metaphysical matter. From this perspective any attempt at modifying the human genome becomes sacrilegious, as an attempt to play God. Such normative assessments may be traced back to positions taken by scientists, especially with regard to genetics. Particularly in the early stages of molecular biology and genetic research scientists tended to treat the genome as an absolute, immutable, and deterministic blueprint of the development of an organism.

In his chapter Peters criticizes both tendencies as attempts to essentalize the genome, in particular the human genome. Like Rehmann-Sutter he points to recent publications of genetic research which describe the dynamic interaction between

[1] Habermas J: Die Zukunft der menschlichen Natur. Auf dem Weg zu einer liberalen Eugenik? Frankfurt, Suhrkamp, 2001.

genome and external influences (epigenetics). He holds that theologians, in particular Christian theologians, should be sensitive regarding tendencies to view human nature solely from a biological perspective (also referred to as *biologization*); rather, they should focus on the holistic concept of the human person lest the Christological base of theological anthropology will be neglected.

Peters cites the chimera question as an example for the problematic aspects of this biologization of human nature. The diversity of chimera categories (human to human; human to nonhuman; nonhuman to human), and even more the inherent quantitative and functional complexities, highlight the medical and ethical problems and contradictions that would result from genetic essentialism and a blanket negation of any kind of chimera research or production. Thus, according to Peters, not polemical or hysterical warnings of playing God, but avoiding bio-idolatry should be the central mandate of theological genethics, particularly in the realm of a Reformed Theo- or Christocentrism.

Jewish, Islamic, and Buddhist GenEthics

Quite distinct from Christian positions, in particular the Roman Catholic position, is Judaism's perspective on genethics. **Ronald M. Green**, a philosophical bioethicist with a Jewish background, alludes to this in his chapter. Judaism is, in general quite liberal with regard to a number of topics, e.g. abortion, in vitro fertilization, prenatal diagnostics, and research involving human stem cells. Not surprisingly, then, Israel has more infertility clinics per capita than any nation in the world. To be sure, debates are also ongoing among Jewish scholars and the public regarding the severity of a child's anticipated congenital abnormalities and/or the threat to the mother's physical or mental health that justify intervention. Still, the climate of discourse is quite different from that among other religions, in particular Christian denominations.

It may be surmised that this openness towards modern forms of medical treatment and diagnosis is peculiar to the secularization of religious perceptions that are found, in particular, among Reform Judaism. According to Green, however, this is not the case. The same liberal attitudes are prevalent among Orthodox Judaism as well. Thus, Green deduces that the foundations for this relatively liberal attitude towards genethical issues are to be found in the religious tradition and literary sources of Judaism itself.

In support of this hypothesis, Green identifies three broad principles as the supporting pillars of Jewish biomedical ethics that are relevant to genetic science: (1) the stringent obligation to preserve human life through medical healing and medical research; (2) a developmental view that accords nascent human life moral weight proportional to gestational age, and (3) Judaism's openness to the application of technological advances, including human biology, for the improvement of human life. Green agrees with other Jewish ethicists that these three ethical principles can be traced back to the Bible and the Talmud.

Following Green's analysis of the respective ethical discourses, it becomes evident that the three pillars are themselves based on three deeper bioethical convictions:

(1) the fundamental ethical value of human reproduction; (2) a rather exclusivist concept of human life as fully developed personal life, paired with a quite restrictive view of the inherent moral value of basic forms of human life such as the fetus (in Judaism, the fetus is not regarded as a person), and (3) the theological concept of human co-creatorship with God, paired with the religious concept of the intentional imperfection of the creation that allows, or even demands, continued human effort towards perfection. Green himself suggests that Jewish bioethics based on these principles could be open for rather progressive forms of human genetic self-modification, including forms of human enhancement.

In many Muslim-dominant countries stem cell research is accelerating rapidly although Islamic ethics of human embryonic stem cell research have yet to reach consensus. This is at least the interpretation of **Siti Nurani Mohamed Nor**. According to her, a perusal of various scholarly Islamic writings suggests that there is a plurality of views regarding the beginning of life, but there is consensus regarding the existence of potential life in the human embryo; accordingly, the Quran declares that the unborn baby has the undeniable right to inheritance. Potentiality, according to Islamic interpretations, does not, however, accord exclusive rights to the embryo, including the right to life. Instead, lawmakers consider every action in light of the choice of the lesser of two evils, in this context foregoing the potential of gene technology vs. infractions of the *maqasid al sharia* (the objectives of Islamic law) which are defined by five basic human interests: life, religion, property, intellect, and family lineage. Islamic scholars find that human embryonic stem cell research does not contravene all five *maqasid.* There is life in these cells but the crime committed against them is minor according to the Sharia. Next, research in this area does not transgress religious barriers because it does not constitute playing God. Third, organ and body parts, including cells and tissues are not considered the property of human beings and consent is, therefore, unnecessary. Fourth, research is highly regarded as extending and protecting the intellect, and seeking knowledge is a noble act in Islam. Finally, research on genetic materials is allowed as long as regulations and controls are laid down to ensure that family lineage is protected within the bounds of legal marriage.

According to **Pinit Ratanakul**, it is – at least in principle – not difficult for Buddhism to cope with new scientific achievements such as genetics and biotechnology. The basic teachings of *kamma* and human nature do not exclude genetic manipulation through gene therapy. The reason for this liberal attitude of Buddhist ethics towards human genetic manipulation, and potentially some forms of human enhancement, is to be found in a general and structural affinity of Buddhism and science, as both see the need for the verification of truth by reason and experience. Beyond this explanation one might assume that it also might have to do with the fact that, in contrast to Western traditions, Buddhism denies the uniqueness of the individual person as a biographical unit of mind and body. However, its teaching of inter-dependence opposes any intervention that has negative social effects. Accordingly, in Buddhist ethical thinking only somatic cell therapy is acceptable, and only for therapeutic purposes, whereas its use, e.g., in cosmetic medicine, for self-enhancement is not. Similarly, germ line cell therapy is ethically questionable due to its potentially negative effects on humanity. The same applies to human reproductive cloning where possible negative effects are considered to outweigh positive ones.

In contrast to this quite accepting attitude towards genetic manipulation in general, Ratanakul argues for a rather restrictive position concerning embryonic stem cell research. This follows from the categorical Buddhist precept to avoid any killing of human life. Despite its open and tolerant spirit the Buddhist position on the preciousness of embryonic life is uncompromising even in the case of excess embryos. According to Ratanakul's interpretation of Buddhist ethics, this precept even outweighs the ethical principle of compassion with such human beings whose quality of life might profit substantially from the results of such research.

Beyond Playing God

To keep up with the rapid progress of modern gene technology religious ethics must engage in new ideas and unorthodox ethical reflections. Denying such reflections would risk rendering religious ethics a mere mouthpiece of anti-technological fears and instincts and a public morality mirroring such attitudes. Avoidance of such attitudes will require diligent efforts to understand relevant forms of biotechnology and their potential risks and benefits, as well as critical self-reflection among religious ethicists.

In the sacred texts of Christianity, Judaism, Islam, and Buddhism, of course no direct reference is made to the ethical dimension of gene technology. Nevertheless, there are concepts of life, principles of ethics, and principles of dealing with the creative potential of humanity in all these scriptures. In addition, all religions have more or less established traditions of sophisticated ethical debate on those subjects and as they pertain to intricate questions of human life. Religious debates, including those on ethical topics, may be treacherous, highly ambiguous, and – at least for outsiders – often irrational. Still, these debates have their own logical framework, and by and large they endeavor to be of public relevance. This does not mean that bioethical debates within religious communities are less intense and rigorous than those in secular environments – quite on the contrary. Indeed, pluralism is not only a characteristic of the modern world but also of the history of religions – religions needed to find ways to deal with pluralism all along.

Despite their wide diversity, all ethical debates among religions are probably based on a common, shared view of humanity as the most tenuous form of life on earth. Human life is itself playing God, without being God. The preservation and reflection of this knowledge may be the very reason why religious ethics is not only useful but perhaps even fundamentally necessary for both individuals' and societies' discourse on gene technologies. This knowledge may indeed be why humankind has adopted religious thought and practice throughout history.

Context, Acknowledgments, and Dedication

This volume goes back to a conference held in Basel on May 22 and 23, 2008, entitled "GenEthics and Religion: Advancing the Public Dialogue on Human Gene Technology and Bioethics as Reflected in Religious Thought" – a joint event organized by the Faculty of Theology of Basel University, the Science and Ethics Advisory Group (SEAG) of F. Hoffmann-La Roche AG, and S. Karger Publishers.

The editors wish to thank the F. Hoffmann-La Roche AG for their generous support of the conference and this volume. They would also like to express their gratitude to the Karger Publishing House, especially to Messrs. Thomas and Steven Karger and Mrs. Gabriela Karger. Their initiative was essential for this enterprise; they served as active members of the organizing committee and accepted this volume for publication. We also thank S. Karger Publishers for untiring support throughout the publication process.

The editors wish to dedicate this volume to the memory of Mr. Steven Karger who passed away several weeks before the conference, on March 5, 2008, at the age of 49 years.

Georg Pfleiderer, Basel
Gabriela Brahier, Basel
Klaus Lindpaintner, Newark, Del.

Pfleiderer G, Brahier G, Lindpaintner K (eds): GenEthics and Religion. Basel, Karger, 2010, pp 12–27

Genes – Cells – Interpretations

What Hermeneutics Can Add to Genetics and to Bioethics

Christoph Rehmann-Sutter

Institute for the History of Medicine and Science Studies, University of Lübeck, Lübeck, Germany

Prologue

I began to write this text when I had to wait a few hours for a flight connection at London Heathrow. I sat down with a coffee and my laptop at a small table, a bit away from the crowds. The topic I was thinking about was how we can make sense of what we know about the genes in our bodies. We know about genes and DNA, about genetic risks associated with mutations we happen to carry, and about the genome functioning within us. All this is something of us, something that belongs to our embodiment, makes it possible for us to live. I listened and looked around.

Complexity is within me, and outside as well. The acoustic space around me was filled by a constant rush from an enormous air ventilation system in the airport hall, an occasional squeak of a not too well lubricated escalator going down to the arrivals level, the chat of some fellow travelers nearby, and a strong, slowly speaking female voice over the loudspeakers giving instructions to two individuals to 'go immediately to gate 14'. Signs everywhere with written instructions for passengers to go here or there, invitations to buy this or that, screens announcing the next flight departures and the corresponding gate numbers. Here, you just need to know where you want to go, to have the right ticket ready, and this immense machine of air transport will interact with you and take you there, I thought. The planes outside seemed like external extensions of the airport machine, little spin-off machines, which fly away and distribute people to places far off around the globe. That complexity outside was clearly a human-made construct, easy to read, *made* to be easy to read. Instructions all around. But our bodies? Do the genes make them readable for us too, testable and foreseeable? Genomes as our bodies' instruction books? That was the question I wanted to think about.

Nobody would doubt that in an airport, the texts we see all around us are texts indeed. The act of gathering information from screens and panels (or from a hastily spoken 'Hi'ere, how're you?' at an immigration officer's desk) is essentially a complex act of understanding, and sometimes, if it is not entirely clear what it means, even interpreting. There is a meaning in all these signs. They really *are* signs, not just things that look like signs. Signs are fascinating constellations. Language is not an object. Language speaks. The information that is understood by passengers making their way through the airport is essentially lingual in its form. Its origin is a human mind, therefore those hearing or reading the messages know that what they hear and see are indeed messages: complex signs that can be understood as meaningful. Somebody wants to say something to you there. But who wants to say something to you through the genome? Is the genetic information information in and 'for' the organism? But in what sense of 'for'?

And what about all those people that I saw sitting, chatting and walking around there in Terminal 5? I know that they are not human-made, and I know that they know it too. My body was not 'built' by my parents and the genome was not 'written' by them. I am glad, otherwise I would perhaps blame them for all the mistakes and limitations within it. It was rather the nature of their bodies that created my body. Our bodies are self-organizing, living organisms and essentially products of nature. How should we comprehend what we explain when we explain their functioning? Biology tells us how cells and bodies function. But then we need to explain to ourselves and to others what it means *for us*. It is only the psychological meaning of an experience that constitutes a 'phenomenon' [1].

We know that scientists who are gathering information about the structure and functioning of human bodies are supposed to avoid all subjective bias. They should measure and observe. The ideal of scientific knowledge is objectivity, provided by impartial measurement and reproducible experiments. But is this division of the world of knowledge into a realm of human information (that can be understood, misunderstood, that sometimes must be interpreted) and a realm of natural information (that can be gathered by measurements and in experiments) not too simple? Interpretation certainly happens also in biology, most active biologists would admit that. Biological theory is not an enumerative process of copying and pasting information from measurements together and adding them up into theories. Such theories would never make sense. To build theories that do make sense and have explanative power is not just an enumerative but a creative act. You need to see a pattern, a picture in all these data. The scientist's task is much more challenging (and rewarding) than just scooping up data and pasting them together: it is putting pieces together in a puzzle without knowing in advance what image the puzzle will have. Sometimes there are several possible combinations leading to different pictures, different 'puzzles' with the same pieces. Theories of the natural sciences sometimes seem to be 'underdetermined' by experimental facts, as Quine [2] has claimed. Applied to genetics: is there an alternative picture of the genome besides the image of the instruction book?

The least we must admit is that the life sciences produce descriptions which, in a certain way, interpret nature. The language of biology is no translation from another language; it describes. And this act of describing is essentially an act of making sense of what relates us to ourselves and to others in a certain way. Biological descriptions interpret the biosphere. They interpret the processes going on in our bodies. And they express their interpretation in a way that is meant to be as objective, as comprehensive, and as accurate as possible, in the form of an understandable language. The life sciences open a space of lingual meanings that are expressed in language, meanings that are 'about' the nature and the functioning of the body with its cells, membranes, proteins and chromosomes. This step of description is therefore, in a very fundamental sense, also a step of understanding. Aristotle, in his treatise *De interpretatione*, spoke of *hermeneia* as the act that is performed *on things* by language [3, p 105]. If this makes sense, hermeneutics, usually thought to be relevant only for the humanities and social sciences, would have relevance also for the natural sciences.

Again these people sitting about around me: what do they know about their bodies, about their genes? Do they have a picture in their mind of how the genes work in their bodies? If they have taken images from the media, their image will probably be an image of DNA as text, or instruction book, or blueprint. The genes would then contain instructions for the body on how to develop, how to stay alive, how to grow, and how to age. This is the message that is abundant in the media. It is the basic picture the media draw of the meaning of the molecular genetic evidence. Scientists themselves helped to create this picture. It corresponded to their expectations of molecular biology, at least in the 20th century [4, 5]. But today the scientific literature, when we listen to researchers, contains a different view and also the media start to take it up [6–8]. The old paradigm of the genome containing the genetic program has come into disrepute and a new idea is emerging. – Now in more detail:

Hermeneutics

The way in which the results of molecular biology are connected with meaningful images and stories, and how it is socialized, politicized and technologically exploited is deeply cultural. Therefore, molecular biology, even if it is a natural science, can fruitfully be looked at from a hermeneutic point of view. The task of hermeneutics, when we follow one of the most important theorists of hermeneutics, Hans-Georg Gadamer, is not merely to develop a procedure to understand difficult texts, but to discover and explore the very conditions under which human understanding takes place [9, Part 2, II.1.c]. When we apply this to molecular biology, this means that the task of (genetic) hermeneutics is to discover and explore the conditions under which the meanings of genes, genomes, genetic texts and genetic factors in human life and societies can be understood.

I want to explore some aspects of this hermeneutic process that manifest themselves in the practical, cultural sides of genetics, mainly those that are connected to the fact that genetics is a heavily *lingual* undertaking. Taking the phenomenon of language seriously in such complex inter-textual relationships (genetic information vs. information about genes), we can more clearly see how cultural understandings of biology and biological theory work together. And we can also deduce what ethical implications different understandings may have.

Political Texts

One of the key moments when genetics became visibly politicized and a political text of genetics was written, was the announcement that a 'working draft' of the human genome sequence was finished. This took place on June 26, 2000; the key figures were President Bill Clinton and Prime Minister Tony Blair. At a media conference in the White House, they announced it as an 'historic achievement' of science. The politicians' task was to find a comprehensible explanation as to *why* it was so important to humankind to know the boring details about the order of the T, A, G, and Cs in their DNA.

According to the White House press communiqué issued one day before, the key point in the subsequent announcement by Clinton and Blair would be:

> 'that the international Human Genome Project and Celera Genomics Corporation have both completed an initial sequencing of the human genome – the genetic blueprint for human beings' [10].

People who did not know about genetics and the biological significance of DNA did possibly know what a 'blueprint' is supposed to be: a detailed plan, more precisely a paper-based reproduction of a construction plan as it is commonly used in architecture or engineering.[1] The image of the blueprint seemed to fit with DNA because it is also a result of a copying process that occurs in every cell division.

However, the message contained in the lightheartedly spoken phrase the 'genetic blueprint for human beings' was complex and contained several elements. We can perhaps better identify them when we tentatively translate 'blueprint' into the term 'construction plan'. I see at least six elements in this statement:

(i) There is a construction plan that contains the details about how we are made;

(ii) Development and life are essentially a constructive process based on a plan;

(iii) The DNA sequence is or represents the construction plan for building the organism, i.e. for its development and life;

(iv) By sequencing the human genome we learn the details of the construction plan for human beings;

(v) Now we know how human beings are constructed, and

(vi) Two countries that are scientific leaders, the US and the UK, have achieved this.

[1] The 'blue' stems from the photochemical reaction leading to Prussian blue that is used in the cyanotype process which was developed by photographer and astronomer Sir John Herschel in 1842. For a century, blueprint was the only available low-cost process for copying drawings [11].

Bill Clinton, in his oral speech, was even more outspoken:

> 'Today's announcement represents more than just an effort making triumph of science and reason. After all, when Galileo discovered he could use the tools of mathematics and mechanics to understand the motion of celestial bodies, he felt, in the words of one imminent researcher, that "he had learned the language in which God created the universe". Today we are learning the language in which God created life' [12].

This breakthrough must be immense, comparable to Galileo's mathematical astronomy and the birth of modern cosmology (that was not anti-religious). The laws of nature were thought of as language that God used in His (perhaps still ongoing) act of creation. With his reference to God and to creation, Clinton *theologized* or *sacralized* the genome. Genome sequences do not just look like language, they *are* language, even the most important language one can imagine: the language in which God created us as living creatures.

It is interesting to see how the old biblical image of the 'book of life' has been transformed by this explanative step. Originally it had a moral meaning. In *Exodus* (32: 32f) it was said that *HaShem* had a book that contains the names of those who are righteous and worthy of life. Those who are blotted out of this book will die. In the context of genomics, the book of life and its language is re-read as God's language, the instructions He uses in His work of creation. The moral meaning, which said that the book of life noted who is worthy and should live, drops away. A new moral meaning is established instead of the old. Now, it is a hopeful source of benefit for human life, of good to be done, particularly of medical innovations and cures for some of the most devastating diseases.

National interests certainly played their roles. Behind the Human Genome Project was an international partnership led by the US and the UK. For this purpose it was useful that it was something so powerful that could be announced: nothing less powerful than the language in which humans were made. It is human life's innermost secret, previously known only and exclusively to the One who created us. Now it would be available and accessible for human use. But it is not just the power that this genetic knowledge provides to those who control it that makes this performance by Clinton and Blair a political text, it is also the structure of this knowledge itself that proves to be power-related. From a Foucauldian perspective it cannot be overlooked that there is a correspondence between the genome as the controller, genetic information that is our 'makeup', the language in which we are created and the power of those who have control over genomic information or who control the experts who provide the necessary genetic know how.

Metaphysical Texts

When the metaphor of the 'program' first appeared in the published literature of molecular biology in the early 1960s, very little was known on a molecular basis about how different genes actually work in development. The genetic program was a story in a nutshell that explained comprehensively in advance how genes could organize development and make life possible. If the genome contains only information (a sequence of four different nucleotides), an explanation was needed of how organisms can be made of information, i.e. how the 1-dimensional sequence of components of the DNA polymer can be transformed into a 3- or 4-dimensional reality of an organism. The program was the key idea.

The notion of the 'gene' originated much earlier, but in another context. It was the problem of inheritance not development that first led biology to talk about genes. Johannsen's famous definition in 1909 [13] aimed at the factors – whatever their precise nature may be – that can be passed on by egg and sperm and determine the traits and characteristics of the next generation of organisms. The gene was an answer to what we can dub the 'bottleneck-problem' of biology [14]. This problem is located within the gap between the generations. How can the form and structure of a species in one generation be rebuilt from the two tiny germ cells that are passed onto the next generation? Oocyte and sperm fuse and build together one single cell, the zygote, which must be capable of developing into the whole new organism. What component of the oocyte can be responsible for this? A major discovery was that it is not the membranes, not cytoplasm, not the proteins, but nucleic acid, essentially the nucleic acid contained in the chromosomes. Chemically it is DNA, a double-stranded, enormously long molecule shaped as a double helix. Rosalind Franklin, James Watson and Francis Crick discovered its chemical structure in 1953. The implication was, however, that a sequence of only four different building blocks, the nucleotides, must be the key responsible element in heredity. How can that be possible? How can a sequence, or the information contained in the sequence, guide development? The bottleneck-problem was tightened from a cellular level to a molecular one.

Information is essentially a pattern of differences that can be reliably translated into another pattern and therefore can be said to determine the second pattern. However, genetic information was not thought to be semantic in the sense of carrying an intention to represent this second pattern. It is neither the previous generation that encodes an intention within its genetic information, nor any spirit in nature, nor God. This is very important to keep in mind. The notion of genetic information was not thought to be intentionalist but purely mechanical.

Before developmental geneticists in the 1980s and subsequently could investigate more and more of the actual mechanisms of developmental processes, and elucidate the precise roles that different genes and gene products play, two of the most productive molecular biologists of the earlier time, François Jacob and Jacques Monod, in a historic paper of 1961, used the term 'program' to explain how the imagined development was possible. The paper was about their discovery of gene regulation in bacteria, the *Lactose operon*. In the concluding section they observe:

> 'The discovery of regulator and operator genes, and of repressive regulation of the activity of structural genes, reveals that the genome contains not only a series of blue-prints, but a co-ordinated program of protein-synthesis and the means of controlling its execution' [15, p 354].

Protein synthesis was the essential process for the development and life of the organism. As we will see later, the crucial part of this quote is the statement that 'the genome contains' the program. DNA contains a program for development, and therefore the cells can execute this program if they are properly equipped. The genome, according to Jacob and Monod [15, p 221], even contained the means of controlling its execution [4, 5]. The DNA molecule was thought of as a central organizer of all essential steps in the life of the organisms. The *Lactose operon* provided an example of how this could work: genes are regulated by regulator molecules that are again synthesized from other genes. The genome is a self-regulatory system controlling the development and behavior of the cell.

In the same year, 1961, another biologist who was more preoccupied with evolution and inheritance than biochemistry, Ernst Mayr [16], published a somewhat similar idea also using the program metaphor. But for him, the metaphor served other needs. His problem was teleology, i.e. the question of how it can be thought possible within a strictly Darwinian framework that allowed for no reference to vital forces, that organisms are such miraculously well-functioning constructions and show goal directed behavior. The assumptions of Modern Synthesis Darwinism combined the molecular evidence of DNA-copying and the 'central dogma' of molecular genetics (information flow goes from nuclear DNA to protein, not back from proteins into DNA) with the Darwinian mechanism of random variation and the survival of the fittest. This excluded the possibility of inheriting phenotypically acquired functions from the previous generation. Everything had to be in the genes. The solution that Mayr saw was given by the idea of a 'genetic program'. If the previous generation passes a program to the next, containing all the information that the system needs to reconstruct itself and to behave, then we do not need any more assumptions to explain an apparent goal-directedness in development and behavior. Programs evolve by chance and selection. Teleology, the obscure old doctrine of nature following certain aims, could be replaced by a cybernetic model of goal-directed, negative feedback systems, or 'teleonomy' [4, pp 196, 221; 16]. The processes *seem* to be goal-directed, but this is only their appearance to us. Actually, development is programmed by DNA information, and it is the program that is passed onto the next generation, perhaps including mutations that may give advantages or disadvantages to the organism and to its chances for reproduction.

This was an argument against vitalism. Vitalism was a basically metaphysical doctrine arguing on the level of ontology. It claimed that living processes contain a nonphysical force or principle that make them what they are, which can never be explained by the physical sciences. Therefore, in Mayr's text, the assumption of the genetic program was also a metaphysical argument, working on the level of ontology. It explained how is it possible that living beings come into existence, and their way of existing in the world. In terms of theory building, the idea of the genetic program replaced intentionality in nature and the vitalist non-physical forces. It was therefore *negative* metaphysics that Mayr was arguing for. With the physically plausible assumption of a genetic program it was no longer necessary to believe in nonphysical forces and intentions to explain the apparent functional organization and goal-directed behavior of organisms. Jacob and Monod [15] were comparatively more constructive in their approach. They saw how such a genetic program could be conceptualized: as a sequence of regulatory steps whereby genes are regulated by the products of other genes.

But was this discursive move to the programs containing genome just a step towards dropping unnecessary metaphysical ballast and therefore in itself ontologically or metaphysically 'innocent' or neutral? I do not think so. In order to see this, we need to advance another 45 years in the history of biology and look at the ideas of contemporary systems biology. Kunihiko Kaneko, a Japanese biologist and influential theorist of complex molecular systems, summarizes the current approach to explaining the emergence of relatively stable and regular developmental pathways of organisms in a very simple way as follows:

> 'Thus the situation is one of mutual influence, not unidirectional causation. Hence, although the genes can be thought of as in some sense controlling such processes, in fact it is not true that an understanding of the genes alone is sufficient for their complete description. For example, even if we were somehow able to obtain the DNA of a dinosaur, unless we also knew the initial conditions of the cellular composition that allow their proper expression of genes, we would not be able to create a Jurassic Park. The conclusion we reach from these considerations is that ... we should be studying models of interactive dynamics. Then, we should inquire whether, within such dynamics, the asymmetric relation between two molecules is generated so that one plays a more controlling role and therefore can be regarded as the bearer of genetic information' [17, p 20].

If we compare this idea that Kaneko is outlining, referring to vast experimental evidence and to mathematical models, with the image inherent in the 'genetic program', several important differences become obvious. The relation between different molecules and processes in the cell are seen as mutual influence, instead of a unidirectional causality contained in sequences of linear *if-then* events. Parts interact in many ways, loops of causal influence going forward and backward, branching in many directions and making the system as a whole relatively open or relatively closed. The division of roles within such a system is not a precondition, but must itself be explained as a result of the interactive dynamics of the system. Therefore, the apparent specialization of DNA as the bearer of genetic information, and the many very important roles singular genes can play in the development of the organism, are products of the interrelation of the parts of the system and their dynamics. This is the second striking difference to the genetic program approach. There, it was thought that a causal program is a precursor of development in the shape of the sequential composition of the DNA polymer. Thirdly, a regularity of developmental events that could be described as something like a program is to be located on the level of those interactive dynamics, not on the level of one component of the system. In terms of the distinction between genotype and phenotype, the assumption of systems biology is that the program (if anybody still wants to talk of programs) is a *phenotypic* regularity. What behaves regularly in foreseeable and reproducible sequences of events is the whole organism within an environment, not one isolated molecule.[2]

Molecular genetics, particularly in the context of developmental biology and genomics, has contributed to this enlargement of the picture. A wide variety of mechanisms that enable the cell to use DNA sequences in different ways have been discovered. The active RNA molecules (still suggestively called 'messenger' RNA) are compiled in much more complicated ways, and most of the RNA molecules (MicroRNAs) are no templates for proteins at all. DNA sequences that code for proteins (containing the genes in a classical sense) and some proteins are multifunctional. Which effects will be realized depends on a multitude of other factors, and sometimes on spatial information as well, i.e. on the place within the cell or a multicellular network. Some genes overlap, some genes can be spliced in multiple ways, depending on the situation, and the resulting RNA variants will lead to different proteins that are all related to the same DNA stretch. Sometimes the cell uses fractions of one and sometimes fractions of the other of the two single DNA strands as a template for producing a functional RNA. There are also switches to alternative reading frames, i.e. the shifting of the three-letter code by one, resulting in different sequence information. There is sometimes even post-transcriptional editing of mRNA, i.e. the introduction of changes in the sequence of an mRNA molecule after its composition, which also leads to a different amino acid sequence of the protein being built from it [14, 19–22]. Genes are multifunctional [23], and therefore they can no longer be considered as independent factors in a chain of events. But this is precisely what the idea of the 'genetic program' suggested.

Taking these and other phenomena into account, Eva Neumann-Held [24] has reconsidered the very concept of the gene from a systems perspective. If we still want to call what explains the biosynthesis of a particular type of protein in a cell a 'gene', we can no longer say that one stretch of DNA is responsible. It is rather a range of factors, interacting with DNA and with each other, and processes sometimes transgressing the boundaries of the cells and the body, that are actually contributing. This set of contributing factors includes DNA, but also much else. Neumann-Held bases her reflection on a groundbreaking book by Susan Oyama from 1985 that has the title 'The Ontogeny of Information' [25] and has inspired many authors to new formulations under the umbrella term of a 'developmental systems approach' [26, 27]. The key idea was a new attempt to theorize development. Previously, we thought development was basically a result of two different information resources, one internal and genetic, the other external, i.e. social, environmental, or cultural. The divide between these two information resources has materialized in the 'nature or nurture' debate in developmental psychology. One school emphasized the genetic contributions (sociobiology, evolutionary psychology, etc.), the other more the social and cultural factors. Oyama's point was that development is better seen as an interaction of both, but not

[2] Elsewhere [18], I have argued that a phenotypic 'program' cannot consistently be said to be a program, because the term program is based on the distinction between prescription and realization.

in the sense of an interaction between two independent types of factors. There can be no genetic factors *without* the environment, and there can be no environmental factors *without* an organism and its internal resources. Both are mutually related, so that we should avoid making a distinction between those two classes of contributing factors. From this, Oyama came to a different understanding of genetic information as developmental information. Developmental information itself has a developmental history. With reference to Gregory Bateson's [28] famous definition of information as 'difference that makes a difference', she explains: Genetic information 'neither preexists its operations nor arises from random disorder. ... Information is a difference that makes a difference, and what it "does" or what it means is thus dependent on what is already in place and what alternatives are being distinguished' [25, p 3].

Genetic information, in the sense of developmental information, is itself 'developmentally contingent in ways that are orderly but not preordained, and if its meaning is dependent on its actual functioning, then many of our ways of thinking about the phenomena of life must be altered' [25]. Following this line of thought, biology's basic picture is being transformed.[3]

The program view assumed that the development of the complex organism we see in the biosphere depends on the existence of genetic information, which can be copied and reproduced from a template. It said that the generations do not transmit a small prototype or the adult structure to the next generation, and no supernatural intentions or forces are necessary for a comprehensive explanation of the development of a new generation. What is transmitted is a list of instructions for making that structure [29, p 2]). The systems view by contrast sees that in fact it is the entire cell that is reproduced, not only some lists of instructions. The cells reproduce not because the genome contains instructions for building it, but

> 'because any inheritance involves passing on DNA and all the cellular and extracellular structures, processes, and materials necessary for its exploitation' [25, p 77].

Information can therefore be seen in at least two different ways: either as something inherent in a pattern that is transmitted or as something that is itself a product of interactive dynamics. Accordingly, two different ontologies of living phenomena are put forward.

But they are not equivalent offers, from which we could choose one arbitrarily. There are today serious reasons for preferring the systemic approach. The most compelling of these comes from science itself.

If information for development is a product of interactive dynamics, it does not exist *before* the development of the system actually takes place. Developmental information continually emerges from the interactive dynamics of the cellular (or multicellular) system and – we need to allow this conclusion as well – continually fades away afterwards. In this sense, genetic information is contingent and ephemeral. DNA is an inert and relatively constant molecule, a source of stability for the system and highly

[3] See our collection [31].

important in many ways, but DNA is *not* the carrier of developmental information. We can say that the organism is essentially a self-informing system. The whole system develops and behaves regularly and predictably in many ways, but the regularity is not a result of a preexisting program.

This theoretical rereading of the molecular evidence in terms of systems cannot take away any of the empirical evidence that we have about the causal involvement of DNA sequences or mutations in the development of certain characteristics like diseases. The systems view, as I understand it, is not an argument against genomics, medical genetics or against DNA-related research in any way. The argument does not work on the level of experiments or empirical work, but on the level of the interpretation of the empirical evidence, which can be seen as plausible or implausible.

Who Is the Author of the Genetic Text?

The impact of the systems approach and of reinterpreting genetic information reaches beyond criticizing the genetic program metaphor. It contains also a critique of other metaphors that are used to explain the meaning of genomic information in terms of *signs* contained in DNA sequences: the book of life, the instruction book, the architecture plan, the blueprint, the text, and related ones, in other words all metaphors that work with a difference between signifier and signified and introduce a semantic relation between DNA as a 'sign for' and the meaning of this sign. Sign metaphors do not work within a systems approach because they presuppose that information for development preexists.

The assumption of preexisting meaning in organisms would also be difficult to defend from a hermeneutic point of view. Signs or compositions of signs (texts) that we use in language are not just prints on paper that can be copied or transformed by certain rules. Because they belong to language, texts are expressions of personal life. In contrast to spoken language, they are permanently fixed expressions of personal life. Language, as Gadamer puts it, is the universal medium of understanding, i.e. a medium in which the acts of understanding itself can occur [9, p 392]. If we have a text, we must assume a writer who transformed meaning into written signs, and there is a reader who transforms written signs back into meaning. Who and where is the writer of the genetic 'text'? The genome is the product of evolution; nobody has 'written' it. Therefore, the text as a metaphor for explaining DNA must always be essentially flawed. It would be a non-written text, part of non-meant language. To seriously claim such a thing implies a conceptual confusion.

Metaphorically speaking, not biologically of course, we could say that in the framework of the systems approach that assumes that genetic information does not preexist development but is itself an emergent product of interactive dynamics, the *body* acts as something like 'the author' of the genetic information. I have myself used this way of speaking [14]. If genetic information as developmental information is composed from step to step as a result of the interactive dynamics in the cells, it is actually the body that brings about this information. This use of the metaphor of the body as author is nonetheless lopsided because the 'bringing about' of the text is certainly no intentional act. It is no act of speaking, but rather mindless, even if our body is not an object but an animated body, this sensible, living body we ourselves are. If we use this metaphor we must see its limits as well.

This holds true for all metaphors. We should not naturalize the metaphors. Often metaphors can be of great help in making sense of a natural phenomenon in a context where sense is not obvious.[4] The metaphor tells a little story, it is a narrative in a nutshell, or, as Paul Ricoeur [31] has said (with reference to Beardsley), a miniature poem. Metaphors graft meaning from one discourse to another and sometimes also modify their meaning when they arrive within the new discursive context. Sabine Maasen and Peter Weingart [32] speak of metaphors as messengers of meaning. The 'author' is such a metaphor of the body in relation to developmental genetic information: a messenger bringing meaning from the discourse about texts into the discourse about embodiment.

To regard the descriptive text as a text is not so far away from what the humanities (as the sciences of the texts) see as a text. Also regarding a historical text, the meaning that we understand is not necessarily the same meaning put into the text by the writer. The act of reading is interpretation. There is a convincing argument by Hans Georg Gadamer that the act of interpretation involves the reader in her or his proper cultural and social context as well. There is no interpretation without the active contribution of the interpreter, who has questions in mind, brings them to the case, questions which might differ considerably from those of the writer or other readers. A text brings something up, raises a subject; but it can only do this because of the *interpreters' participation* [9, p 391]. The situation of the biologist who constructs a text in her or his attempt to understand is therefore not so different from that of the historian or the reader of historical literature who interprets a text as a testimony that has been written under different circumstances. But mind that this way of speaking about texts in genetics treats the *interpretation* of DNA as a text, not the sequence of DNA.

To see description as interpretation can perhaps also help to liberate our minds from old images and doctrines. We can *play* with metaphors more easily when we are aware of what they do, with which questions they resonate, and what they do not do. Living in highly technologized worlds where most things are constructed according to plans and instructions are abundant, it is no surprise that biologists started from the question how the organism is constructed according to instructions or a plan. Today we obviously need new metaphors for the genome-cell relationship that make better sense but are still easy to grasp. Many things are possible.

We could, for example, see the genome like a library. A library is – in contrast to a book – not an organized text, not a message with a stable content. The meaning that

[4] Another line of critique against the genetic program view would be that it naturalizes its metaphor. The assumption is that programs are not just our 'way of understanding' but that genes 'really are' programs, i.e. that they 'work as' programs.

readers take out of a library depends on which books they are picking, which books they are leaving behind, what proportion of the books they check out they are reading, in which order they are reading them, what they read, hear or see from elsewhere, and on what they make of the texts they are using. The library metaphor is, however, also lopsided, because libraries still contain books that were written by authors, whereas the genome is not a written text at all. But when we put the emphasis on the side of the reading act, the comparison is more accurate. The cell 'reads' its genome like readers read when they are selectively and creatively using a library.

Or we can compare the mechanisms of how words gain meaning in spoken or written language. There is a similar phenomenon that we know well in language to that of the multifunctionality of genes or proteins. Words have multiple meanings; they are polysemous. *Polysemy* means the capacity for a sign to have various meanings. How can we understand each other with words if the words can mean different things? The answer is that one word can say different things, but which meaning is in the foreground depends on the context in which it is used.

When we now look back on the history of the genetic program we see that it was an attractive preconception about the meaning of the genome drawing on the language of computers, whose attractiveness can be explained in the historical and cultural context in the second half of the 20th century. It was essentially an anticipated story of how the genes work, invented before experimental knowledge in developmental genetics was available. Lily Kay has written the 'history of the genetic code' in a book with the ambiguous title 'Who Wrote the Book of Life?' [4].

The book of life was meant to be a book written by nature. But it turns out that it is rather written by humans, scientists in particular, but not just by them. We have seen politicians play their role as well. This writing, it emerges, happens not on the level of the genes but on the level of the explanations that have been promulgated and were selected because they seemed to be more meaningful than others.

When scientists today explain genetics to the general public there is a risk that they use the same old preconceptions over and over again, just because they feel or anticipate that this is what the public 'out there' will understand. They might be right that it is the program view and the family of sign metaphors that they do know 'out there'. But this is nothing other than what, a few years earlier, the public was told by scientists to believe. I believe that the storyteller (in the best sense of the word: the scientist truly explaining genetics to the general public) has a responsibility to be not just rhetorical but authentic.

We Do Our Genes

It is easy to see that genome interpretations do have ethical implications. Space allows me only to mention some. Consider pre-symptomatic genetic tests. Within the framework of program genomics, a mutation in a cancer-related gene is understood as information that the body carries and that instructs the cells to make tumors, or, in the case of mutated tumor suppressor genes, as a 'fault' in the information that tells the body how to suppress tumor cells that might emerge. In the framework of systems genomics it looks different. The same mutation in a cancer-related gene is an indicator for an elevated likelihood that the body under certain circumstances can get cancer, or an indicator that the body's capacities to prevent cancer are weakened. Within this interpretation, the mutation does not indicate 'information for making cancer' or a 'fault in the information on how to suppress cancer'. There is no information 'for cancer' in the body right now. The patient does not walk away with the idea that 'I have a fault in me', but knows about a risky, even dangerous condition and hopefully about possible precautions to take.[5] The practical measures that are recommendable will not be different, but the relationship of the patient to his or her body will be different. This has implications for the language used in genetic counseling, and also for the ideas of what 'good counseling' means.

Or consider embryo ethics. One argument that plays a crucial role in the politics and ethics of embryo research (stem cells, cloning, etc.) is potentiality [33]. Many bioethicists who defend this argument base it on the genome. The embryo, after fusion of the egg and sperm, carries the full genome for the individual. Alfonso Gómez-Lobo [34] writes:

> 'The potentiality to become a male or female human adult is due to the biological program contained in the genome'.

And therefore, he argues, it is not the gametes that are entitled to moral and legal protection but the embryo and the fetus, right from its beginning. Or read Otfried Höffe who argues in a similar way:

> 'Was Kritiker als "blossen Zellhaufen" abtun wollen, trägt von Anfang an, als befruchtete Eizelle mit doppeltem Chromosomensatz, das volle genetische Programm für die Entwicklung eines Menschen in sich. Das Programm liegt tatsächlich rundum vor, in seiner notwendigen und zureichenden Gestalt' [35, p 137].[6]

In the framework of systems, obviously, this argument does not work. To recognize this not necessarily changes our ethical attitudes of care and parental responsibility, which we owe to embryos. It will not make them freely accessible for research, because there are other grounds for responsibility as well. But nevertheless, it will change the reasons why philosophers can defend a responsible legal framework for embryo research. And other positions might be less dogmatic and allow more freedom for individual ethical judgment.[7]

[5] Thanks to Lorraine Cowley, herself a genetic counselor, now social researcher in genetics, who brought to my attention that the usual language in genetic counseling sessions in Britain includes 'genetic fault' for mutations and 'instruction book' for the genome.
[6] 'What critics want to dispatch as 'just a heap of cells' in fact carries the whole genetic program for the development of a human being, from its very beginning as a fertilized egg cell. The program is integrally present, in its necessary and sufficient form'.
[7] Implications for morally interpreting the role of the embryo donor in stem cell research are pointed out by Scully and Rehmann-Sutter [36].

Bioethics, I want to conclude, needs a contextual hermeneutics of the body, of genes, of cellular systems, and of nature, which is methodically and topically not separate from the hermeneutics of traditions, art, discourses and morality. This bioethics that society needs cannot be a shallow kind of 'applied ethics' where a certain moral dogma is taken for granted and 'applied' to a practical dilemma. Society needs a bioethics that is part of broad practical philosophy and works closely with the social sciences and humanities. Philosophy needs to go into the world and to work in interdisciplinary collaborations with social and cultural studies. It also needs to learn from people out there in the situations, who live those new kinds of dilemmas, who feel where and how moral questions arise. In the practical circumstances surrounding biotechnologies or biomedicine, the deepest questions about what is 'good', what is our desire for 'a good and fulfilled life with others and for the others' in a social and natural environment arise. Questions about what are 'just institutions', and what is 'ethical governance' in bio-societies arise. Bioethics therefore also needs a strong theoretical component reflecting independently the needs of society with regard to the conceptual tools and methods used. In this respect, biosciences are a fruitful soil on which ethics can grow.

References

1 Giorgi A, Giorgi B: Phenomenology; in Smith JA (ed): Qualitative Psychology. A Practical Guide to Research Methods. London, Sage, 2003, pp 25–50.
2 Quine WV: On empirically equivalent systems of the world. Erkenntnis 1975;9:313–328.
3 Ricoeur P: The Course of Recognition. Cambridge, Harvard University Press, 2005.
4 Kay LE: Who Wrote the Book of Life? A History of the Genetic Code. Stanford, Stanford University Press, 2000.
5 Keller EF: Decoding the genetic program. Or, some circular logic in the logic of circularity; in Beurton P, Falk R, Rheinberger H-J (eds): The Concept of the Gene in Development and Evolution. Historical and Epistemological Perspectives. Cambridge, Cambridge University Press, 2000, pp 159–177.
6 Angier N: Scientists and Philosophers Find that 'Gene' Has a Multitude of Meanings. New York Times, 11 November 2008.
7 Zimmer C: Now: The Rest of the Genome. New York Times, 11 November 2008.
8 Nerlich B, Hellsten I: Genomics: shifts in metaphorical landscape between 2000 and 2003. New Genet Soc 2004;23:255–268.
9 Gadamer H-G: Wahrheit und Methode. Grundzüge einer philosophischen Hermeneutik, ed 5. Tübingen, Mohr, 1986.
10 White House Office of Press Secretary: President Clinton Announces the Completion of the First Survey of the Entire Human Genome, June 25, 2000. www.ornl.gov/sci/techresources/Human_Genome/project/clinton1.shtml; accessed March 2, 2010.
11 Ware M: Cyanotype. The History, Science and Art of Photographic Printing in Prussian Blue. London, Science Museum, National Museum of Photography, Film, and Television, 1999.
12 http://www.genome.gov/10001356; accessed March 2, 2010.
13 Beurton P: Genbegriffe; in Krohs U, Toepfer G (eds): Philosophie der Biologie. Frankfurt, Suhrkamp, 2005, pp 195–211.
14 Rehmann-Sutter, C: Genetics, embodiment, and identity; in Grunwald A, Gutmann M, Neumann-Held EM (eds): On Human Nature. Anthropological, Biological, and Philosophical Foundations. Berlin, Springer, 2002, pp 23–50.
15 Jacob F, Monod J: Genetic regulatory mechanisms in the synthesis of proteins. J Mol Biol 1961:3:318–356.
16 Mayr E: Cause and effect in biology. Science 1961; 134:1501–1506.
17 Kaneko K: Life: An Introduction to Complex Systems Biology. Berlin, Springer, 2006.
18 Rehmann-Sutter C: Instruierte Reproduktion. François Jacobs Konzeptionen des genetischen Programms 1961 bis 1997. Figurationen Gender Literatur Kultur 2003;4:29–48.

19 Stotz K, Bostanci A, Griffiths P: Tracking the shift to 'postgenomics'. Commun Genet 2005;9:190–196.
20 Rehmann-Sutter C: Genes in labs. Concepts of development and the standard environment. Philos Nat 2006;43:49–73.
21 Rehmann-Sutter C: Genetics, a practical anthropology; in Düwell M, Rehmann-Sutter C, Mieth D (eds): The Contingent Nature of Life. Bioethics and the Limits of Human Existence. Dordrecht, Springer, 2008, pp 37–52.
22 Alberts B, Johnson A, Lewis J, Raff M: Molecular Biology of the Cell, ed 5. New York, Garland, 2008.
23 Fogle T: The dissolution of protein coding genes in molecular biology; in Beurton PJ, Falk R, Rheinberger H-J (eds): The Concept of the Gene in Development and Evolution. Historical and Epistemological Perspectives. Cambridge, Cambridge University Press, 2000, pp 3–25.
24 Neumann-Held EM: The gene is dead – long live the gene: conceptualizing genes the constructionist way; in Koslowski P (ed): Sociobioogy and Economics. Berlin, Springer, 1999, pp 105–137.
25 Oyama S: The Ontogeny of Information. Developmental Systems and Evolution, ed 2. Cambridge, Cambridge University Press, 1985.
26 Oyama S, Griffiths PE, Gray RD (eds): Cycles of Contingency. Developmental Systems and Evolution. Cambridge, MIT Press, 2001.
27 Stotz K: Organismen als Entwicklungssysteme; in Krohs U, Toepfer G (eds): Philosophie der Biologie. Frankfurt, Suhrkamp, 2005, pp 125–143.
28 Bateson G: Steps toward an Ecology of Mind. New York, Ballantine, 1972.
29 Smith JM, Szathmáry E: The Origins of Life. From the Birth of Life to the Origin of Language. Oxford, Oxford University Press, 1999.
30 Ricoeur P: Metaphor and the main problem of hermeneutics. New Literary History 1974/75;6:95–110.
31 Neumann-Held EM, Rehmann-Sutter C (eds): Genes in Development. Re-Reading the Molecular Paradigm. Durham, Duke University Press, 2006.
32 Maasen S, Weingart P: Metaphors and the Dynamics of Knowledge. London, Routledge, 2000.
33 Deckers J: An analysis of the arguments underpinning UK embryonic stem cell legislation on the embryo's status. J Stem Cells 2007;2:47–62.
34 Gómez-Lobo A: Does respect for embryos entail respect for gametes? Theor Med 2004;25:199–208.
35 Höffe O: Menschenwürde als ethisches Prinzip; in Höffe O, Honnefelder L, Isensee J, Kirchhof P (eds): Gentechnik und Menschenwürde. Cologne, DuMont, 2002, pp 111–141.
36 Scully JL, Rehmann-Sutter C: Creating donors: the 2005 Swiss law on donation of 'spare' embryos to hESC research. Bioethical Inquiry 2006;3:81–93.

Christoph Rehmann-Sutter
Institute for the History of Medicine and Science Studies
University of Lübeck
Königsstrasse 42, DE–23552 Lübeck (Germany)
Tel. +49 451 7079 9810, E-Mail rehmann@imgwf.uni-luebeck.de

Pfleiderer G, Brahier G, Lindpaintner K (eds): GenEthics and Religion. Basel, Karger, 2010, pp 28–39

Controversies about Human Dignity: Implications for Biotechnology

James F. Childress

Institute for Practical Ethics and Public Life, University of Virginia, Charlottesville, Va., USA

Introduction

This essay examines appeals to human dignity (HD) – respect for human dignity – in public bioethics, that is, in efforts to formulate or recommend public policies on bioethical matters. HD has figured prominently in bioethical debates in European and other international contexts, particularly but not only when human rights have also been invoked. It has also been invoked in bioethical debates in the US about such issues as human reproductive cloning. In the US it has also been a rallying cry for resistance to a variety of biotechnologies, especially among evangelical Protestants, Roman Catholics, and neoconservatives – many of whom are now labeled 'theocons' [1].

In general, philosophers in the analytic tradition, dominant in Anglo-American philosophy, tend to be more suspicious of appeals to HD than philosophers in the Kantian or phenomenological traditions. Ethicists within particular religious traditions – or philosophical ethicists attuned to those traditions – tend to be more sympathetic to appeals to HD, along with ethicists who draw heavily on the international human rights tradition.

Until the announcement of Dolly's birth, HD was mainly invoked in bioethical debates in the US around 'death with dignity'. Both sides in the debate about of whether physician-assisted suicide should be decriminalized appealed – and continue to appeal – to 'dignity' [2]. The protean nature of HD, so evident in that debate, provides one reason for objection to appeals to HD – some charge that it is 'a loose cannon' [3], while others charge that its use is 'inflationary' [4]. I concur: in public bioethics, at least in the US, HD is loosely used, overused, and overextended. As a result, it functions more to obfuscate than to illuminate bioethical debates.

Still others go farther. Philosopher Ruth Macklin [5] contends that '[d]ignity is a *useless concept* in medical ethics and can be eliminated without any loss of content' (italics added). The important and defensible content of HD, she argues, can be found in clearer concepts of respect for persons or – even better – respect for autonomy. An even sharper criticism is that appeals to HD actually damage the quality of public discourse about biotechnologies.

One of the worst recent offenders, according to many critics of appeals to HD, is the President's Council on Bioethics (hereafter PCB). In 2002, it issued a report entitled 'Human Cloning and Human Dignity' that inexplicably left the idea of human dignity unanalyzed and its connections with human cloning undeveloped [6]. There are suspicions that the PCB invoked HD, without analyzing it, in order to smuggle in religious and theological concepts without having to acknowledge them as such. In 2008, in a volume entitled 'Human Dignity and Bioethics: Essays Commissioned by the President's Council on Bioethics' (March 2008), PCB members joined by others return to HD and examine its many dimensions in a set of essays that provides further evidence of the religious/theological context of the PCB's invocation of HD [7; for a sharp critique, see 8].

In order to explore different conceptions of human dignity in public bioethics, I want to draw a distinction between (1) HD as a standard of duties and rights, and (2) HD as a standard of virtue[1]. I will illustrate the logic of these two conceptions by reference to debates about human reproductive cloning and human genetic enhancement, before arguing, in conclusion, for a precautionary process – not a precautionary principle – to address concerns about biotechnology's threats to HD.

Conception I: HD as a Standard of Duties or Rights

HD often appears as a standard of action in accord with various duties that should be followed and/or rights that should be respected. This first conception of HD presupposes some property, characteristic, or capacity of human beings or persons that is the basis for dignity and respect for dignity.

There are variations in Conception I about whether HD is a foundational standard or one standard among others. This is evident in different international human rights documents. These variations also appear in bioethical frameworks. For instance, some European formulations of bioethical principles include dignity alongside others, such as autonomy, dignity, integrity, vulnerability, and solidarity [10]. In these formulations, the principle of HD functions as one principle alongside others. Specifying the content of HD would involve determining, for example, what exactly a principle of HD adds to principles of autonomy and integrity.

[1] In a recent discussion, Doris Schroeder [9] identifies four concepts of dignity: (1) Kantian, (2) aristocratic, (3) comportment, and (4) meritorious dignity (influenced by Aristotelian thought). The first and the fourth are closer to the conceptions developed in this essay, but even the first and fourth are too narrow for our purposes, while the second and third focus exclusively on societal rank and societal expectations. For an analysis that has influenced my discussion, see the work of Beyleveld and Brownsword [3].

By contrast, Pope John Paul II's encyclical 'Evangelium vitae' offered a conception of HD, understood as 'transcendent dignity... as the visible image of the invisible God', as foundational: it is the source of an order of 'human rights inherent in every person and prior to any Constitution and state legislation' [11]. Here, specifying the content of HD requires determining those rights and their correlative duties.

One important question, as already suggested, concerns the relation between HD and respect for autonomy. Macklin [5] seems to think that whatever is ethically important in HD can be expressed in respect for autonomy (or respect for persons). Still others [12] sharply distinguish respect for autonomy from HD and stress the tension between them. Certainly in some contexts, HD is offered as a justification for restricting rather than allowing voluntary actions. This justification appeals to an objective standard of HD over autonomy. Examples appear in legal decisions that voluntary actions, such as a dwarf's decision to allow himself to be tossed in bars or a woman's decision to perform in a peep show, can be overridden by concerns about HD [3, 13]. These cases suggest that tension exists not only between HD and autonomy, but also between the two conceptions of HD that I am distinguishing here.

Obviously, this first conception rules out violations of specific basic duties and rights, derived from or backed by HD. However, appeals to HD attempt to capture something more than particular violations – this 'something more' is often viewed as *indignity*, *insult*, or *affront*. Under dignity-denying conditions, human actors may lose their *sense* of dignity but their dignity is not lost, or even reduced, because it is a characteristic or property of the human being or person [14]. Several Roman Catholic thinkers have expressed this point: using reproductive cloning (or other reproductive technologies) *violates* a child's rights and his/her HD, but it does not *diminish* or *destroy* that child's dignity – the child produced through cloning, for example, would still have HD, as a child created in God's image, despite the process of cloning [15].

In addition to debates about how to specify HD in various duties and rights, there are debates about its range and scope (e.g., the status of the embryo) and about its weight, strength, or stringency (e.g., whether it is absolute or prima facie), as well as about how to justify this standard and specific duties and rights, for instance, in philosophical or theological anthropology.

To illustrate how this first conception of HD works, let us consider some common arguments about human reproductive cloning. Several international organizations, such as UNESCO and the WHO, and several individual thinkers appeal directly to HD, rather than to HD through specific human rights, to argue for a ban on human reproductive cloning [16]. But it is not easy to argue from HD to an absolute ban on human reproductive cloning. Consider, for instance, the common argument against human reproductive cloning based on HD and the specific derivative duty not to treat human beings merely as means, instruments, or objects, rather than as ends in themselves. The problem with such an instrumentalization argument against human reproductive cloning is that it proves too little or too much.

It is hard to make the case that human reproductive cloning necessarily involves the kind of instrumentalization that HD excludes. Human reproductive cloning does not necessarily involve treating offspring merely as means or instruments to other ends. Instead, human reproductive cloning could involve the range of (often mixed) motivations and intentions that mark both natural procreation and assisted reproduction such as in vitro fertilization. Reasons for having a child through sexual intercourse frequently view the child, at least in part, as a means, for instance, to save a marriage or to extend the family line. Using reproductive cloning to have a child one of whose kidneys or bone marrow or cord blood would be a good match for a dying sibling or to have another child to replace a dying child – these might seem to be particularly problematic examples of instrumentalization, but they may too be mixed. Nor is it clear that other bad effects, relative to HD, would necessarily result through human reproductive cloning – these include the denial of individuality, autonomy, and equal dignity.

It would be easier to make the case for a ban of human reproductive cloning based on HD as a standard of duties and rights, *if* we could develop 'an operational definition of human dignity' along with explicit arguments about how reproductive cloning infringes HD [17]. We would need to determine exactly how human reproductive cloning infringes HD, in terms of both specific rights and the remainder expressed by HD.

There are strong reasons to support a ban on human reproductive cloning that I do not address here – I am here limiting my attention to HD and related conceptions.

Conception II: HD as a Standard of Virtue

A second conception of HD has a different focus. For instance, in reprogenetics, this conception focuses less on what these technologies do to embryos and more on what the uses of these technologies display about the character – the virtues or vices – of the users or providers. This is a vision of HD as a standard of virtue, excellence, and flourishing. It dominates the bioethical pronouncements of Leon Kass, both in his own work over several decades and in his role as founding chair of the President's Council on Bioethics, and it pervades the PCB's reports.

From this perspective, HD is part of a 'richer bioethics' than is found in common bioethical principles. This conception of HD focuses on violations that go beyond the infringements of rights and duties, however important these may be. In particular, Kass [18] argues that certain acts, practices, and policies are wrong even if they violate no particular rights or duties.

The main language of violation of, or threats to, HD on this conception features *dehumanization*, *degradation*, and *debasement*, all realized in the post-human future depicted in Aldous Huxley's 'Brave New World', a book Kass [18] returns to again and again because he believes its fictional world is increasingly similar to our own. The flaw in the Brave New World is not the lack of freedom, as Huxley seemed to think, but a deeper defect – difficult and perhaps impossible to diagnose through conventional bioethical principles such as respect for autonomy – the voluntary violation of standards of human excellence, what Kass [18] calls 'willing dehumanization'.

While we easily recognize certain violations of principles such as respect for autonomy – for example, in the failure to get voluntary informed consent for research participation – 'we are slow to recognize threats to human dignity, to the ways of doing and feeling and being in the world that make human life rich, deep and fulfilling' [18]. For Kass, that tardiness is unfortunate because the major threats from the new biotechnologies are to HD, not to our principles of liberty and equality.

Take the example of human reproductive cloning again: from the standpoint of HD as a standard of virtue, opponents of human reproductive cloning focus primarily on the cloners, those seeking cloning, and the society that allows it. Without neglecting the impact on the children created this way, they tend to emphasize the actors. In this account, instrumentalization is condemned more as a display of bad or deficient character than as a violation of duties and rights (unless duties to self are recognized).

Central to this second account of HD is the 'wisdom of repugnance'. Kass [19] observes that many people find the prospect of human cloning 'offensive', 'grotesque', 'revolting', 'repulsive', and 'repugnant'. He concedes that 'revulsion is not an argument' and that past repugnancies have sometimes been overcome, for better and for worse. However, 'in crucial cases', according to Kass [19], repugnance is 'the emotional expression of deep wisdom'. Examples of wise repugnance include bestiality, cannibalism, mutilation of a corpse, and raping or murdering another human being. Repugnance is sufficient to condemn these actions, without rational justification – '[t]he repugnance at human cloning belongs in this category' [19].

A few critical points. First, despite what Kass seems to suppose, these examples vary greatly. For instance, even though cannibalism evokes horror, it could conceivably be ethically justifiable under some horrible circumstances that do not involve killing but only eating human flesh following someone's death. And certainly mutilation of the corpse, which is broad enough to include removal of organs for transplantation, can be ethically justified under some circumstances. Revulsion may be a starting point, but sometimes it can and should be overcome.

Second, even though Kass provides several arguments, beyond repugnance, for banning human reproductive cloning, the 'wisdom of repugnance' appears to be sufficient for him. He assumes that repugnance is universal in response to the above cases and that it is a natural expression (though it can be lost) rather than a cultural product. Still, in disputes, what counts is the repugnance of the 'wise', that is, those who are 'morally serious' [18, 19]. The wisdom of repugnance appears to be the repugnance of the 'wise'. This argument is problematically circular.

Third, appeals to sentiments, even among the putatively wise and morally serious – however they are identified – are still problematic. Disgust cannot reliably discriminate among actions in morally relevant ways. It may instead reflect mere prejudice or moral blindness [12]. Psychologist Steven Pinker's attack on HD, in response to the PCB's work, has the title 'The stupidity of dignity' and the subtitle 'Conservative bioethics' latest, most dangerous ploy' [8]. As the title suggests, the article attacks repugnance, revulsion, etc., as an inappropriate basis for bioethical policy. Not only does this conception of HD, as a standard of virtue, run the risk of prejudice or blindness, it tends to offer indiscriminate criticism rather than precise guidance for discriminating judgments about different biotechnologies.

Finally, another problem is whether and how to justify restrictions of liberty on the basis of sentiments evoked by violations of HD as a standard of virtue, where there is no infringement of HD as a standard of duties and rights. This is particularly difficult in a liberal, pluralistic democracy, with its various standards of virtue.

Conflicting Appeals to HD in Public Bioethics: The Example of Human Genetic Enhancement

Both conceptions of HD have been invoked to assess and often to oppose a variety of developments in biotechnology. I have already used human reproductive cloning in passing to illustrate different conceptions of HD. Now I will offer some observations about human genetic modification, particularly genetic enhancement, in light of these two conceptions of HD.

As many recognize, the distinction between therapy and enhancement is problematic, indefensible, and unworkable. It does not provide a bright line, in part because it relies on contested concepts, such as disease, health, and normal state, with unclear boundaries. Nevertheless, for now, we can ask: for genetic and other interventions usually labeled enhancements, what are the major ethical and social problems? Different concerns emerge in two contexts: individuals choosing for themselves versus choosing for their offspring. Ordinary ethical issues in both contexts concern the balance of benefits and risks – with particular attention to safety; fairness and equality; and voluntary informed choices.

For instance, one ethical concern is that when enhancement interventions are available, only the wealthy will be able to afford them and, as a result, the gaps between the genetically enhanced and unenhanced will produce new or augment old social stratifications. Of course, we should also worry about other social inequalities that also have an impact, and perhaps an even greater impact on individuals' lives – for example, their living environment, their nutrition, and their family's resources.

Even if it is not feasible to address all these inequalities, the question remains: what, if anything, should be done to prevent further inequalities based on access to genetic technologies for enhancement. For instance, should we try to avoid these inequalities by preventing some research and development of genetic technologies? That would be ethically problematic, in part because many enhancements use technologies developed in the first instance for important and indisputably therapeutic purposes and goals. Allen Buchanan and colleagues go farther, suggesting that certain genetic enhancements may be justifiable and perhaps even obligatory, as long as there is a fair distribution of the benefits and burdens [20].

In 'Beyond Therapy' [21], the PCB devotes brief, almost perfunctory, attention to probable inequalities in access to genetic enhancements, but it also notes that prices for enhancement biotechnologies can be expected to decline over time and allow greater access by people with fewer resources. The PCB views this 'familiar' issue of equality and justice as more 'socially important' than ethically significant. The 'more fundamental ethical questions about taking biotechnology "beyond therapy" concern not equality of access, but the goodness or badness of the things being offered and the wisdom of pursuing our purposes by such means' [21].

The PCB believes, as Leon Kass does, that there are sources of 'disquiet' about biotechnological improvements that go well beyond familiar bioethical principles. This 'disquiet' erupts in an 'initial revulsion':

> 'If there are essential reasons to be concerned about these activities and where they may lead us, we sense that it may have something to do with challenges to what is naturally human, what is humanly dignified, and to attitudes that show proper respect for what is naturally and dignifiedly human' [21].

This passage clearly expresses the second notion of HD – a standard of excellence and flourishing, a standard of virtue. The PCB specifies this conception of HD through the several points, which apply, in varying degrees, to different biotechnologies aimed at happiness or fulfillment.

- Appreciation of and respect for 'the naturally given', threatened by hubris
- Dignity of human activity, threatened by 'unnatural' means
- Preservation of identity, threatened by efforts at self-transformation
- Full human flourishing, threatened by spurious or shallow substitutes [21].

The PCB's conception of human nature, flourishing, and dignity is not the only plausible one. Some other conceptions feature human rationality, creativity, and the like. Some even hold that we are most human when we are making and using technologies. Related conceptions focus on human self-creation and self-modification. According to Neil Levy, in 'Neuroethics',

> 'We are … animals of a peculiar nature: we are self-creating and self-modifying animals. We alter our own minds, and use technological means to do so. This is not something new about us, here and now in the "postmodern" West (for all that so much about us, here and now, *is* genuinely new). That is the kind of animal we human beings are' [22].

While the PCB and other critics contend that certain uses of biotechnologies for enhancement will lead to a 'post-human' future that transcends human nature and violates human dignity, 'transhumanists'

> 'see human and posthuman dignity as compatible and complementary. They insist that dignity, in its modern sense, consists in what we are and what we have the potential to become, not in our pedigree or our causal origin. What we are is not a function solely of our DNA but also of our technological and social context. Human nature in this broader sense is dynamic, partially human-made, and improvable' [23].

Similar points apply to opposition to genetic enhancements based on the first conception of HD, connected with human rights – they too tend to appeal to a more or less static human nature in interpreting basic rights. But there are also important differences in their arguments. Opposition to genetic enhancement, based on the first conception of HD, tends to focus on our breaches of our duties toward – or our infringements of the rights of – children and future generations. For instance, drawing on both human rights and human equality, based on HD, George Annas argues that genetic enhancements, particularly through germ-line interventions – just as human reproductive cloning – constitute a new kind of 'crime against humanity', parallel to terrorist acts [24]. He is particularly concerned about the creation of new species of humans, perhaps post-humans, and, as a consequence, interspecies conflict and warfare, with genocide being an inevitable outcome [25]. This vision is even darker than the one presented by the PCB, but Annas fails to make a compelling case for the probability of this dark scenario [26]. He invokes speculative possibilities.

The two conceptions of HD I have identified – there may be others – appeal to HD and draw its ethical implications in distinctive ways – however much overlap exists. Nevertheless, as we have seen, there are problems in major arguments for coercive bans against the development and use of certain biotechnologies – here I used human reproductive cloning and human genetic enhancements as examples – based on either of these conceptions of HD. I note again that other kinds of arguments may be stronger than those based on HD.

A Public Precautionary Process

Nevertheless, we need to ask how, in light of widespread concerns about threats to HD, we should assess and respond to biotechnologies, particularly genetic technologies that could be used for enhancement? Assessment will include attention to possible benefits as well as possible problems.

Prospective technology assessment is complicated. Historical perspectives on previous technologies may make us skeptical of such prospective assessments, calls for which may be more common in the area of genetic interventions than other technologies.

In an essay entitled 'Technology Assessment from the Stance of a Medieval Historian', Lynn White Jr. insists that technology assessment requires 'cultural analysis' because a technology's impact is filtered through a society's culture, broadly understood [27]. It involves attention to the 'imponderables', not only to easily measured factors. Among his several case studies is one of alcohol, which was distilled from wine as a pharmaceutical at Salerno, the site of one of Europe's most famous medical schools. How, he asks, could anyone have provided a helpful prospective assessment of alcohol in the 12th century? Alcohol was praised in medieval literature as a pharmaceutical that was effective for chronic headaches, stomach trouble, cancer, arthritis, sterility, falling or graying hair, and bad breath. It was also supposed to be good for people who had a 'cold temperament'. But widespread drunkenness and urban disorder soon became problems. To shorten the history, we now face major problems of alcohol-related traffic deaths, cirrhosis of the liver, and the like. As White observes, 'A study group eight centuries ago, equipped with entire foresight, would have failed at an assessment of alcohol as we today fail' [27].

For many, several current and possible genetic and other biotechnologies pose a threat to important social values, including HD. For most, there are at least reasons for concern – political philosopher Michael Sandel notes that genetic and other enhancement technologies are disquieting, unsettling, and troubling, and cause unease and 'moral hesitation' [28]. Suppose, however, we lack adequate ways to determine the *magnitude* of possible negative outcomes, in part because of different and conflicting conceptions of HD and other values, or to determine the *probabilities* of negative outcomes, because of uncertain cause-effect relations and 'imponderables', in part because biotechnologies are filtered through complex social and cultural beliefs, values, and practices.

Our ordinary discourse embodies several maxims of precaution, such as 'better safe than sorry'. Precaution involves taking care and using protective measures in advance of negative events. These maxims of precaution have sometimes been elevated into a *precautionary principle.*[2] Several versions of this principle focus on appropriate societal actions in light of threats of *serious* or *irreversible* damage, even in the absence of full scientific certainty, while other versions expand the principle's scope beyond serious and/or irreversible harm.

A general precautionary principle cannot provide coherent guidance because there are many different threats, hazards, and uncertain risks, and efforts to avoid any single one must also attend to others. More specifically, the failure to develop some biotechnologies may create risks just as the failure to stop or control those biotechnologies. Given the risks on all sides, several argue, the precautionary principle 'leads in no direction at all' [30, 32]. This point is important, for example, because *enhancement* possibilities often emerge from the development of technologies with *therapeutic* possibilities. But refraining from developing such therapeutic technologies would leave persons with diseases and other medical conditions at risk.

Nevertheless, in some contexts, depending on what is valued and what is at risk, it may be ethically justifiable and even obligatory for society to take some precautionary steps, despite insufficient scientific evidence, to avoid or reduce a threat or hazard where the harm would possibly be both serious and irreversible. Some critics even accept a 'modest version of the precautionary principle', focused on the avoidance of catastrophe [30], or an 'anti-catastrophe principle' [33].

[2] Explicit formulations, defenses, and applications of the precautionary principle date from the 1970s. For interpretations and at least cautious defenses of the precautionary principle, see United Nations Educational, Scientific and Cultural Organization (UNESCO), The Precautionary Principle [29]. For several critical perspectives on the precautionary principle, see [30–32].

In the wake of uncertain, even non-catastrophic threats, some precautionary measures may nevertheless be justified. Some societal interventions may reduce the possibility of the harm, while others may reduce the severity or magnitude of that harm. Rather than all or nothing – either permit or ban certain biotechnologies – it may be justifiable to employ intermediate strategies for non-catastrophic threats, while seeking additional information through research and other means and also establishing ways to monitor and reconsider these intermediate strategies as additional information becomes available.

What we need is a *public precautionary process* that can address some of the perceived threats of biotechnologies to HD and other values. This process is more defensible and helpful than a precautionary principle. It would represent an appropriate stance of (a) ethical fallibility and (b) imperfect information in light of the unpredictability of consequences. The process should be transparent and should involve the public as well as scientific experts in deliberation about appropriate policies in response to perceived threats, for example, to HD. In a democratic society, it is important to develop risk-avoidance or risk-reduction policies that are generally consistent with the society's basic values and the public's reflective preferences. The acceptability of any precautionary measures will depend, to a great extent, on various social, cultural, and psychological perspectives. If we accept a *public precautionary process*, we may be able to develop nuanced and discriminating societal judgments and adopt appropriate precautionary measures.

It may be premature to try to set long-range policies, but it is not premature to undertake a serious and sustained *public precautionary process* in light of concerns about HD. And, in view of the range of concerns, even if sometimes overblown, some modest precautionary measures could be warranted – these could include educational programs, professional self-regulation, and limited, targeted governmental interventions, ranging from public hearings to regulation and even prohibition in the face of highly probable serious harms and wrongs.

What is the role of HD in this precautionary process, especially since many worry that it is at risk from several biotechnologies? Some ethicists recommend purging HD from our moral-social-political vocabulary. I do not go that far. However, as this essay argues, discourse appealing to HD needs to be clearer and more rigorous – it is not enough merely to invoke HD.

First, it is important to distinguish two conceptions of HD: HD as a standard of rights and duties, and HD as a standard of virtue. In my judgment, this is a matter of transparency and integrity in public discourse. Rhetorically, appeals to HD gain considerable force from duty-based and rights-based approaches, for instance in the Kantian and human rights traditions, but in fact they often reflect HD as a standard of virtue. However, HD as a standard of virtue is a weaker basis for coercive laws and policies because of competing visions of human excellence and flourishing. Nevertheless, it is quite legitimate for religious and other groups to articulate their visions about HD in public bioethics. This may include, for instance, efforts to broaden the ordinary scope of HD to include embryos and the human species. It may also include visions of human excellence and flourishing beyond human rights/duties. What is less defensible is the too common appeal to HD as a trump card without the kind of clarity, specification, and argumentation needed. This failure provides fodder to critics who contend that appeals to HD are useless and others who view it as dangerous.

This first step is a crucial one, because it points the direction for further analysis. On the one hand, if HD as a standard of rights and duties is at stake, then it would be possible to undertake a careful examination to determine which policy (for instance, a ban or regulation or permission of human reproductive cloning or of human genetic enhancement) would best protect the rights and duties of all the relevant stakeholders as well as the underlying value of HD if it can be specified in the particular context.

On the other hand, if the second conception is involved, then it will be necessary to specify exactly how certain biotechnologies violate HD by dehumanizing and degrading those who use them or the societies that tolerate their uses, and to make a stronger argument for why, in a liberal, pluralistic society, this particular conception of HD as a standard of virtue should be enforced.

References

1 Linker D: The Theocons: Secular America under Siege. New York, Anchor, 2007.

2 Gentzler J: What is a death with dignity? J Med Philos 2003;28:461–487.

3 Beyleveld D, Brownsword R: Human Dignity in Bioethics and Biolaw. Oxford, Oxford University Press, 2001.

4 Birnbacher D: Ambiguities in the concept of menschenwurde; in Bayertz K (ed): Sanctity of Life and Human Dignity. Amsterdam, Kluwer Academic, 1996, pp 107–121.

5 Macklin R: Dignity is a useless concept. BMJ 2003; 327:1419–1420.

6 President's Council on Bioethics: Human Cloning and Human Dignity. New York, Public Affairs Press, 2002. http://bioethics.gov/reports/cloningreport/index.html.

7 President's Council on Bioethics: Human Dignity and Bioethics: Essays Commissioned by the President's Council on Bioethics. Washington, President's Council on Bioethics, 2008. http://www.bioethics.gov/reports/human_dignity/index.html.

8 Pinker S: The stupidity of dignity: conservative bioethics latest, most dangerous ploy. The New Republic, May 28, 2008.

9 Schroeder D: Dignity: two riddles and four concepts. Camb Q Healthc Ethics 2008;17:230–238.

10 Hayry M: European values in bioethics: why, what, and how to be used? Theor Med Bioeth 2003;24:199–214.

11 Pope John Paul II: Evangelium Vitae. Rome, Vatican, 1995. http://www.vatican.va/edocs/ENG0141/_INDEX.HTM.

12 Kuhse H: Is there a tension between autonomy and dignity?; in Kemp P, Rendtorff J, Johansen NM (eds): Bioethics and Biolaw. Vol II: Four Ethical Principles. Copenhagen, Rhodos International Science and Art Publishers/Center for Ethics and Law, 2000, pp 61–74.

13 Malby S: Human dignity and human reproductive cloning. Health Human Rights 2002;6:103–135.

14 Munzer SR: An uneasy case against property rights in body parts; in Paul ER, Miller FD, Paul J (eds): Property Rights. Cambridge, Cambridge University Press, 1994, pp 259–286.

15 National Bioethics Advisory Commission: Cloning Human Beings. Vol. I: Report and Recommendations. Rockville, National Bioethics Advisory Commission, 1997. http://www.georgetown.edu/research/nrcbl/nbac/pubs.html

16 UNESCO: Universal Declaration on Bioethics and Human Rights. Adopted 19 October 2005.

17 Caulfield T: Human cloning laws, human dignity and the poverty of the policy making dialogue. BMC Med Ethics 2003;4:E3.

18 Kass LR: Life, Liberty and the Defense of Dignity: The Challenge for Bioethics. San Francisco, Encounter Books, 2002.
19 Kass LR: The Wisdom of Repugnance. The New Republic, June 2, 1997; reprinted in Pence GE (ed): Flesh of My Flesh: The Ethics of Cloning Humans: A Reader. Lanham, Rowman & Littlefield, 1998, pp 13–37.
20 Buchanan A, Brock DW, Daniels N, Wikler D: From Chance to Choice: Genetics and Justice. Cambridge, Cambridge University Press, 2000.
21 President's Council on Bioethics: Beyond Therapy: Biotechnology and the Pursuit of Happiness. Washington, President's Council on Bioethics, 2003. http://www.bioethics.gov/reports/beyondtherapy/index.html.
22 Levy N: Neuroethics: Challenges for the 21st Century. Cambridge, Cambridge University Press, 2007.
23 Bostrom N: In defense of posthuman dignity. Bioethics 2005;19:202–214.
24 Annas G: American Bioethics: Crossing Human Rights and Health Law Boundaries. New York, Oxford University Press, 2005.
25 Annas G: Genism, Racism, and the Prospect of Genetic Genocide. Prepared for Presentation at UNESCO 21st Century Talks 1994. http://www.geneticsandsociety.org/article.php?id = 1994.
26 Fenton E: Genetic enhancement – a threat to human rights. Bioethics 2008;22:1–7.
27 White L Jr: Technology assessment from the stance of a medieval historian; in Medieval Religion and Technology: Collected Essays. Berkeley, University of California Press, 1978, pp 261–276.
28 Sandel M: The Case Against Perfection. Cambridge, Harvard University Press, 2007.
29 United Nations Educational, Scientific and Cultural Organization (UNESCO): The Precautionary Principle. 2005. http://unesdoc.unesco.org/images/001395/139578ee.pdf
30 Sunstein C: Laws of Fear: Beyond the Precautionary Principle. Cambridge, Cambridge University Press, 2005.
31 Holm S, Harris J: Precautionary principle stifles discovery. Nature 1999;400:398.
32 Engelhardt HT Jr, Jotterand F: The precautionary principle: a dialectical reconsideration. J Med Philos 2004;29:301–312.
33 Posner R: Catastrophe: Risk and Response. New York, Oxford University Press, 2004.

James F. Childress
Department of Religious Studies, 148 Amphitheater Way, University of Virginia
PO Box 400126
Charlottesville, VA 22904–4126 (USA)
Tel. +1 434 924 6568, Fax +1 434 924 1467, E-Mail jfc7c@virginia.edu

Pfleiderer G, Brahier G, Lindpaintner K (eds): GenEthics and Religion. Basel, Karger, 2010, pp 40–51

Global Bioethics, Theology, and Human Genetic Engineering: The Challenge of Refashioning Human Nature in the Face of Moral and Religious Pluralism

H. Tristram Engelhardt, Jr.

Department of Philosophy, Rice University, Houston, Tex., USA

There Is No Substantive Secular Global Bioethics to Guide Human Genetic Engineering

The possibility of human genetic engineering underscores our lack of agreement regarding morality, human flourishing, and the meaning of human existence. In particular, it is impossible to justify a single, common canonical and content-full bioethics as an authority to guide human genetic engineering. The impossibility is grounded in a number of substantive impediments. First, our societies are marked by the culture wars [1], struggles among the partisans of different moral, religious, and metaphysical views as to which view should constitute the dominant or leading culture (i.e., the *Leitkultur*), and what the content should be for a bioethics of human genetic engineering. Moreover, there is no basis to hold that all secular moral disagreements can be set aside by philosophical reflection and sound rational argument, in that the disputing parties often do not agree about the basic premises and rules of evidence that lie at the foundations of their disagreements. As a consequence, the contentions are interminable. Even with regard to secular moral concerns, we are destined to live in the culture wars, including strident disagreements regarding the use of human genetic technologies, including human genetic engineering.[1] This state of affairs is

[1] Bioethics as a secular undertaking does not have the resources to set our moral pluralism aside. At the theoretical level, bioethics promised the secular equivalent of a theology, a moral vision that could supply the basis for cultural orientation. At the level of practice, it promised the secular equivalent of a cadre of priests and chaplains who were also culturally ordained (i.e., authorized) to provide bioethical consultation, serve on ethics committees, etc. [2]. The problem is that, given the intractable character of our moral diversity, bioethics as a secular field cannot provide the guidance both sought and initially expected. Given our moral diversity, clinical bioethics provides relatively little actual bioethics in the

often obscured by paradoxically passionate proclamations of consensus in the face of persistent disagreements. There are assertions of a common morality, including supposedly universally accepted lists of indubitable human rights, evidence contrary to their supposed indubitability notwithstanding. Matters are further complicated by the profound moral and metaphysical gulf separating those who recognize that God exists and those who deny His existence, as well as the gulf separating the various religions. In closing, I explore what Orthodox Christianity knows to be appropriate for a bioethics of human genetic engineering in order to lay out a religious understanding in principle sympathetic to genetic engineering, as well as further to underscore the moral religious diversity characterizing the human condition.

The Plurality of Moralities

If one means by a morality a generally coherent set of settled judgments about what it is to act rightly, about what the good is we should pursue, and about what it means to be virtuous as well as to have a good character, we do not share one morality, even as a secular morality. Moralities can be different, even if these moralities share the same values but rank them differently. Common moral values do not constitute a common morality. For example, different rankings of liberty, equality, prosperity, and security will justify either a social-democratic morality and polity such as in Germany or an elitist, capitalist polity, such as in Singapore. Differences will also be a function of disparate views regarding right-making conditions. Moralities are different when they justify discordant views about the cardinal elements of human life, such as when it is obligatory, permitted, or forbidden to take human life, tell lies, have sex, redistribute property, and/or engineer human biological nature. These differences are reflected in disagreements regarding such issues as the moral propriety of abortion, homosexual acts, redistributive taxation policies, social-welfare states, capital punishment, physician-assisted suicide, euthanasia, and the genetic engineering of human biological nature.[2] There is a plurality of moralities and therefore a plurality of applied ethics, including genethics.

This moral diversity constitutes an impediment to discovering a single, content-full, secular, global bioethics and therefore a common substantive genethics. Moreover, the diversity cannot be set aside through sound rational argument. Agrippa, a 3rd-century philosopher, observed that there are five reasons to hold that philosophical argument will not resolve foundational moral disputes: (1) after 700 years, no

sense of normative guidance [3, 4]. That is, clinical bioethicists relatively infrequently offer normative moral guidance, usually offering instead legal advice, conflict mediation, and clarification of concepts and analysis of arguments. As with the secular revolution of the Enlightenment, so, too, with regard to bioethics, the hopes for a uniformity of moral vision have gone aground on the intractability of moral pluralism.

2 There is a significant diversity of views regarding human genetic engineering [5–11].

one had succeeded by means of sound rational argument to resolve conclusively the disputes at hand; (2) disputants lodge their arguments within their own perspectives and therefore argue past each other. Moreover, absent commonly justified basic premises and rules of evidence, disputants (3) argue in a circle, (4) beg the question, or (5) engage in an infinite regress [12, 13]. As a consequence, morality remains irreducibly plural, and therefore bioethics, including genethics, is irreducibly plural. Different bioethics are grounded in different moral, religious, and metaphysical views, and a final choice among them cannot be made on the basis of a secular sound rational argument. Given this circumstance, one would expect the disputatious character of the moral discourse that we in fact encounter. Moral diversity is substantive and manifest [14].

It is in light of this diversity that one can appreciate why there will not be moral agreement as to whether

1 human biological nature should be maintained as we find it and that no human genetic engineering should be allowed;
2 human biological nature should generally be maintained as we find it, and human genetic engineering should be used only to correct diseases and defects, but not to enhance human nature beyond the general character of humans as we now find them (although the goals and boundaries between treatment and enhancement will be significantly disputed)[3];
3 human biological nature as we now find it may be enhanced, although the enhancement should be constrained by certain substantive binding norms (about which there will be considerable dispute);
4 human biological nature may be enhanced as long as one proceeds prudently with the consent of the actual persons involved and without malevolence to future persons. In the last case, one is left with a range of acceptable futures. Absent knowing what God requires, man is the measure of all things, and we do not share a common canonical view of morality and human flourishing. The possibility of human genetic engineering discloses the diversity of views regarding human flourishing.

Recognizing our moral diversity does not entail affirming a moral relativism or denying that there is a canonical moral truth. The character of our moral pluralism does show that secular moral reflection is unable to determine the nature of that truth. That is, one may be forced to embrace a secular moral epistemological skepticism but not a metaphysical moral skepticism [16]. In any event, one finds a plurality of bioethics of human genetic engineering, ranging from those which enthusiastically endorse human genetic enhancement, such as John Harris [10], to those which would wish to forbid these interventions, such as Jürgen Habermas [9]. Intractable moral controversy will characterize genethics. Genethics is plural.

[3] For an exploration of the constraints of secular moral reflection on the human genetic technologies, see Engelhardt [15].

The Illusion of Consensus

One might still hope to deny this conclusion regarding the salience of moral pluralism by observing that, if there is in reality such moral diversity, why do so many reasonable people talk about the possibility or even claim the reality of consensus? There are at least six features of the sociology and politics of moral claims that account for the obscuring of moral diversity and the creation of an illusion of consensus, if one means by consensus a common agreement regarding some moral matter, not simply the predominance of a particular moral viewpoint or the dominance of a political or social coalition among persons who share strategic political and social interests.

1. There is often a sampling bias on the part of those who proclaim consensus – those with whom many work and live tend to be like-minded, thus obscuring the depth of actual moral diversity.
2. There is the pressure of political correctness, which tends to hide diversity, in that people tend not to voice views openly that will provoke a strong condemnation from those dominating the public forum.
3. There is paradigm-grounded blindness to moral diversity, given the false consciousness a paradigm can support. Because all are guided by the paradigms they embrace, they tend to discount the positions and viewpoints of opponents through regarding them more as noise than as information. Nor are those with whom they disagree, along with their positions, included in the wide reflective equilibria of those who proclaim a consensus or the existence of an overlapping consensus.
4. There is also the force of the strategic agendas that underlie the creation of ethics commissions and committees, which help to create an illusion of general consensus – usually, those establishing ethics committees do not appoint persons with significantly divergent moral views because the result would be endless debates and not the production of recommendations, especially not the recommendations that those creating the commission would find acceptable.
5. There is the self-interest of those who hold a particular moral position which induces them to ignore moral and religious difference – this discounting of moral difference advantages those who wish to sell their services as moralists by allowing them to claim that they know, can explain, and can defend *the* moral viewpoint of reasonable persons, rather than to acknowledge that they are advocates of a particular moral or philosophical sect shaped by a particular sense of, and presuppositions regarding, the morally reasonable.
6. There are political advantages to advancing one's views as rationally canonical so that one is warranted in ignoring or marginalizing the views of one's opponents as irrational. That is, by claiming that one's moral views are self-evidently right, one may succeed in establishing one's particular agenda, such as having a particular list of human rights accepted without subjecting those rights' claims to the usual critical political processes of controversy, compromise, and amendment (e.g., leading persons falsely to conclude that certain moral claims are self-evident). These

various factors in different ways both create the illusion of consensus and render the illusion of consensus attractive. Claims of consensus, though unfounded, are explicable.

Finally, even if there were per impossibile a consensus in the sense of full agreement on a moral issue, the question would then arise as to what would follow morally from such a consensus and why. That is, even if a consensus were to exist, its capacity to identify moral truth would remain unestablished, because one must first show that the consensus is well founded in terms of the merit of the claims advanced, independent of the consensus itself. After all, the history of moral views demonstrates that different views are preponderant at different times in human history. The consensus of a particular majority is of no self-evident moral significance. In short, one must first establish what follows morally, why, and when, from the mere fact of the matter that many agree to a particular moral fashion.

Separated by Heresy and Disbelief

The plurality of secular moral visions and views regarding the permissibility of human genetic engineering is compounded by the plurality of religious moral visions, as well as by the starkness of the gulf separating the moral life-world of committed theists and that of committed agnostic (not to mention atheist) secularists. The contrast between the life-world and metaphysical commitments of those who recognize the universe as issuing from the creative act of God, versus those who approach the universe as if it were ultimately meaningless, as if it came from nowhere, went nowhere, and for no ultimate purpose, will differ regarding the significance of human biological nature and its morally permissible alterability. Those who know that God exists, and who know that He is someone like traditional Christians know Him to be, experience everything differently from committed secularists, in that they recognize that the world is not surd, but the creation of a very particular personal God. For those who recognize all being as grounded in the creative act of God, all of created reality can thus be recognized deo-centrically as united properly towards one focus. This is acknowledged by Immanuel Kant (1724–1804), as when he affirms that acting as if God existed provides the basis for approaching reality with the heuristic expectation of deep unities in nature [see, for example, 'The Critique of Pure Reason' A672–675 = B700–703]. In terms of this point of focus, one can then begin to ask about the proper use of human genetic engineering in light of what humans are ultimately meant to be. In contrast, the agnostic secularist who does not even accept Kant's as-if God must regard all as coming from nowhere, going nowhere, and for no discernible ultimate purpose. Within this perspective, there is nothing humans are really meant to be. In this perspective, humans are left with creating moral narratives guided by different visions of human flourishing, so that their accounts of human genetic engineering will always be plural.

Also and quite importantly, absent rightly recognizing God, morality is itself importantly transformed in its significance: there is no longer a necessary unity of the genesis, justification, and motivation for morality or a clear rational ground for the claims of morality always trumping those of prudent rationality. This state of affairs led Kant, who was likely an agnostic if not an atheist,[4] to assert that the absolute claims of morality are in the end incoherent, absent at least acting as if God existed. As Kant puts it, it is 'our duty to promote the highest good; and it is not merely our privilege but a necessity connected with duty as a requisite to presuppose the possibility of this highest good. This presupposition is made only under the condition of the existence of God, and this condition inseparably connects this supposition with duty. Therefore, it is morally necessary to assume the existence of God' [18]. Anscombe (1919–2001) is even more critical of morality absent God: 'It is as if the notion "criminal" were to remain when criminal law and criminal courts had been abolished and forgotten' [19, p 6]. Absent God, not just as a point of religious orientation, but even as a point of metaphysical orientation, morality is changed in its character. Of all the disagreements separating moralities, the cleft itself between those who recognize God's existence and those who approach the world without knowledge of God is the most profound.[5] Absent any point of ultimate orientation, and given our intractable moral pluralism, common moral constraints on human genetic engineering can at best be rather sparse, such as (1) acting prudently so as not to cause more harm than good (however goods and harms should be calculated and compared), while (2) not using any actually existing persons without their consent, or (3) acting with malevolence towards future persons.

Simply turning to religion does not resolve the challenges of moral plurality, although the existence of God warrants the affirmation of a point of ultimate orientation. Those who recognize the existence of God are for their part divided, for not all are in agreement as to Who exactly God is, much less what He requires of us. This is importantly true with respect to Christians, for Christians are divided by heresy. Orthodoxy Christians, Roman Catholics, and the various Protestant denominations are separated by substantive differences of dogma that generally make each one a heretic to the other. For example, the Roman church, through councils it holds to be ecumenical and through papal declarations, has pronounced anathemas on those who deny papal universal jurisdiction, papal infallibility, the *filioque*, the immaculate conception, purgatory, and indulgences, and on those who affirm the legitimacy of second marriages after divorce on grounds of *porneia*. All of this should merit

[4] As Manfred Kuehn indicates in his biography of Kant, 'Kant did not really believe in God' [17].

[5] The stark cleft between secular and religious viewpoints has not been without bloodshed. The French Revolution and the October Revolutions, which in great measure were secular revolts against Christianity, led eventually to the killing of tens of millions in the name of secular understandings of justice, fairness, and human rights [20, 21]. The animal rights movement offers a further example of a contemporary secular social-moral phenomenon that has been associated with violence against property and lives [22].

for Orthodox Christians the status of excommunicated heretics. Indeed, the Roman church, the Orthodox, and the Protestants are separated in their understanding of the meaning of Church, in that the ecclesiology of the Orthodox, the Roman Catholics, and the Protestants is foundationally disparate. For example, the first rejects the universal jurisdiction and infallibility of any bishop, while among the Protestants there is generally a loss of a recognition of the continuity of doctrine and of the episcopacy as necessary characteristics marking the one, holy, catholic, and apostolic Church. As significantly, there are foundationally different appreciations of the nature of theology itself,[6] as well as the norms for proper conduct.[7]

These differences involve disparate views of philosophy's role in theological reflection. For example, Pope Benedict XVI in his dialogue with Habermas exemplified the central role of philosophical rationality in Roman Catholic theology when he argues that, in order to avoid the '*pathologies in religion* that are extremely dangerous … it [is] necessary to see the divine light of reason as a "controlling organ". Religion must continually allow itself to be purified and structured by reason…' [26]. The result is that for Roman Catholicism philosophy becomes a cardinal third thing between God and man. Philosophy becomes the mediator between God and man. That is, the relation between God and man is mediated by the requirements of philosophy in general and of moral philosophy in particular. The result is, as Benedict XVI envisions, the claims of theology would then be placed within the governance of philosophical rationality. What Benedict may not appreciate is that this leads to a recasting of God in the image and likeness of human rationality, which is plural. Human rationality reflects the canons of philosophy of a particular age. God is in the process humanized by such theologies and becomes a cultural expression.

In contrast, the Orthodox know that the living God transcends our categories of thought so that, to quote David Bradshaw's account of the cleft between the East and the West, the Orthodox appreciate that there can be 'no concept of God ... [and that] God [is] not an essence to be grasped intellectually, but a personal reality known through His acts, and above all by oneself sharing in those acts' [27]. Moreover,

[6] The Fifth Council of Constantinople, fully equivalent to the Ninth Ecumenical Council (1341, 1357, 1351), endorsed the traditional Orthodox Christian appreciation that theology is primarily noetic, not discursively rational, as Roman Catholics generally hold. For an account of the differences between Roman Catholic and Orthodox theology, see Engelhardt [23].

[7] As an example of the moral differences separating Roman Catholic and Orthodox Christian morality consider the issue of intentionally telling a falsehood so as to deceive. Orthodox Christians affirm St. John Chrysostom's (AD 347–407) injunction: 'The straightforward man does great harm to those he will not deceive' [24]. As St. John Cassian the Just Roman (ca. AD 360–435) states, 'Holy men and those most approved by God employed lying, so as not only to incur no guilt of sin from it, but even to attain the greatest goodness' [25] In addition to the quite distinct Orthodox Christian view with regard to the telling of falsehoods, Orthodox Christians know that sins are not just voluntary but also involuntary and of ignorance, unlike the West, which holds that a necessary condition for an act to be a sin is that the act is done voluntarily and with knowledge (there must be a mens rea). Further, Orthodox Christians have no doctrine of double effect as was fabricated in Western Christianity so as to distinguish between direct and indirect abortions. Christians approach bioethics from quite different foundations [16].

theology in the strict sense is not undertaken primarily through a life of academic study, but most properly through a life of asceticism, which supports the possibility of noetic knowledge, an immediate experience of God [28]. As a consequence, some of the greatest Orthodox theologians of the 20th century never completed college [29]. These differences establish quite different bases for theologically determining the constraints on human genetic engineering. This leads inter alia to disparate appreciations of where to find a normative magisterium that can provide guidance regarding the ethics of human genetic engineering bearing on such issues as human germ-line engineering. This is the case, for Roman Catholics, Orthodox, and Protestants are separated by what one with intellectual honesty ought to recognize as a separation over substantive matters of doctrine, making each heretical to the other. Religious and moral diversity is real. Moreover, there are within this plurality religious and secular moral perspectives that consider themselves as uniquely canonical.

Having given grounds for abandoning the search for a moral and religious consensus in bioethics in general and with regard to substantive moral constraints on the use of human genetic engineering in particular, let me further illustrate this pluralism by outlining what Orthodox Christian theology offers. This account underscores an element of our moral diversity, while also showing that secular moral-epistemological skepticism does not entail a metaphysical-moral skepticism. The answers to the moral questions that confront us do not lie simply in an embrace of a particular religious tradition, but in recognizing and achieving a form of theological knowledge that brings experience of a rightly ordered relationship with God, Who is the truth. Orthodox Christianity appreciates that truth is in its foundations not a set of propositions, but personal, the Trinity, and that the proper content of morality is found in our relationship with the Trinity. Answers with regard to the appropriate use of human genetic technologies will then be a function of understanding what uses are compatible with approaching the Trinity.

Orthodox Christian views regarding human genetic engineering are quite positive because the Orthodox view of medicine is positive, and human genetic engineering is a part of medicine. As St. Basil makes clear in his Answer to Question 55 in the 'Long Rules': 'Each of the arts is God's gift to us, remedying the deficiencies of nature...' [30, p 330]. The use of medicine is in general blessed. Indeed, we are expected to develop and use medicine. In particular, St. Basil recognizes the use of medicine as a legitimate response to the adverse health consequences of Adam's sin. 'Just as we would have no need of the farmer's labor and toil if we were living amid the delights of paradise, so also we would not require the medical art for relief if we were immune to disease, as was the case, by God's gift, at the time of Creation before the Fall' [30, p 331]. Medicine (and therefore human genetic engineering) may properly set aside as far as possible the physical infirmities that are the results of sin. As St. Basil puts it, 'When we were commanded to return to the earth whence we had been taken and were united with the pain-ridden flesh doomed to destruction because of sin and, for the same reason, also subject to disease, the medical art was given to us to relieve the sick, in some degree at least' [30, p 331]. All of this is to say that the use of medicine and surgery has been positively affirmed in Orthodox Christianity.

This approach to medicine and the biomedical sciences should regard human genetic engineering as permissible, indeed laudable, at least when set within certain important constraints. For example, were it possible to use human genetic engineering to incapacitate a gene that causes a serious disease such as retinoblastoma, and even remove it from the germ-line so as to protect subsequent generations, there appears to be no prohibition against doing so, and even in general an obligation to do so. This is the case in particular because human genetic engineering can in this case be regarded as a special instance of micro-surgery or micro-level medicine. One would be removing a gene that reflects a consequence of the Fall. In addition, following St. Basil, the reference point for normal human health is Eden before the Fall. There is thus much treatment that might appear to many as enhancement that should in fact properly be regarded as therapeutic and therefore permissible. Indeed, Methuselah after the Fall lived to the age of 968 years [Gen 5:27].

There are, though, norms that do impose constraints on the use of human genetic technologies. These would exclude anything (1) that would set aside the character of humans as male and female, in that the Orthodox are gender-essentialists in recognizing humans as created from the beginning as male and female [Gen 1:27, 5:2; Matt 19:4]; (2) that would transform human biological nature so that our embodiment would no longer be the embodiment taken on by Christ, in which we all should share, or (3) that would render humans so diverse so as not to be cross-fertile and therefore no longer one species that can together increase and multiply [Gen 1:28]. Beyond that, (4) the use of human genetic engineering should not become an all-consuming undertaking. As St. Basil notes, 'Whatever requires an undue amount of thought or trouble or involves a large expenditure of effort and causes our whole life to revolve, as it were, around solicitude for the flesh must be avoided by Christians' [30, p 331]. That is, the use of human genetic engineering, like all other human projects, must be radically relativized through reference to the ascetic pursuit of the kingdom of heaven. Finally, the use of human genetic engineering must be constrained by other norms such as those that prohibit the destruction of human embryonic life. An Orthodox Christian assessment of human genetic engineering recognizes a very particular context within which to understand what is permissible. The use of human genetic engineering must be compatible with a Christ-centered pursuit of salvation.

The Orthodox Christian approach to the morality of using medicine and the genetic technologies underscores the circumstance that there are no moral answers from 'nowhere'. This is the case because any substantive morality cum bioethics that can supply content-full answers cannot be disengaged from a very particular way of life and from particular stances regarding the meaning of humanity and reality. A substantive morality not only defines orthodoxy but also, and very crucially, orthopraxy. A substantive morality offers particular ways of thinking about, feeling about, and acting in the world. As a consequence, communal boundaries are necessary and it is appropriate as well as inevitable that differences will be underscored. For this reason, it is important to note the positive role of excommunication both on grounds of heterodoxy and when imposed as a temporary disciplinary prohibition because of a failure of orthopraxy (e.g., when a person does not adequately fast, pray, and/or frequently confess his sins and thus separates himself from the Communion chalice until orthopraxy is restored). Excommunication nourishes the integrity of a community. Excommunication by drawing lines teaches what it means to be fully in community, to be in communion, to see, feel, and act in the world from the fullness of a particular somewhere. Such line-drawing is crucial because only within a substantive 'somewhere' can one share a coherent common understanding of the moral life and of human flourishing.

A content-full moral vision at once creates and strengthens moral focus while establishing boundaries and points of separation. When one recognizes and lives the fact that the universe comes from the creative act of a very particular, personal God with Whom all are called to become united, one will live in ways that will contrast with the moral norms and concerns of those who do not recognize this point of orientation. Views about the morality of human genetic engineering, because they are a function of our judgments about the significance of being human and the character of human flourishing, will reflect the moral pluralism that fuels the battles in the culture wars.

Always a View from Somewhere

Moral pluralism is not just a function of religious pluralism. Secular moral aspirations to affirm a canonical, content-full, albeit anonymous set of values turn out to shatter into a plurality of secular moral perspectives. Secular moral reflection has no advantage over the religious in terms of offering a liberation from the cacophony of moral pluralism. There is no way out of the moral pluralism that characterizes the human condition other than to find a unique and ultimate point of meaning and truth. For example, Orthodox Christianity must always remind us that morality is not just about values, but about responding to a very particular call to union with a very particular and very personal God, the God Who is the Trinity. In such circumstances, one will not be able to reason from outside Orthodox Christianity to the norms all should follow, but will need first through right worship and right belief to enter into the rightly directed pursuit of holiness to which all are called. All substantive accounts of morality and bioethics, including the bioethics of human germ-line engineering, are views framed from within the particular perspective of a particular somewhere (i.e., a view that presupposes a particular moral sense, thin theory of the good, understanding of moral rationality, etc.). The various insights of Orthodox Christianity cannot be disconnected from Orthodoxy as a whole. It is only within that whole that these insights have their defining content and force.

We will find ourselves divided as to the proper use of our emerging powers of human genetic engineering. The human condition is one characterized by moral disagreement, which are only underscored when one attempts to answer the question as to how one ought to remake human biological nature through genetic engineering. The attempt to establish a single, content-full, global bioethics for the use of human genetic technologies not only goes aground on moral diversity but is intensified. Because there is no general or common secular substantive bioethics or general religious understanding that can uncontroversially guide the use of human genetic engineering. Our disagreements are rendered more salient when focused on the very recoding of our biological future. We confront different views as to whether and to what extent human genetic engineering supports or is an impediment to the moral life and flourishing of humans. Illusions of consensus to the contrary notwithstanding, we are divided by moral and religious pluralism. Our disagreements will not go away.

References

1 Hunter JD: Culture Wars: The Struggle to Define America. New York, Basic Books, 1991.

2 Engelhardt HT Jr: The ordination of bioethicists as secular moral experts. Soc Philos Policy 2002;19:59–82.

3 Rasmussen L: Clinical bioethics: analysis of a practice; unpublished doctoral dissertation. Houston, Rice University, 2003.

4 Engelhardt HT Jr: The bioethics consultant: giving moral advice in the midst of moral controversy. HEC Forum 2003;15.4:362–382.

5 Bayertz K (ed): GenEthics: Technological Intervention in Human Reproduction as Philosophical Problem. Cambridge, Cambridge University Press, 1994.

6 Buchanan A, Brock D, Daniels N, Wikler D: From Chance to Choice: Genetics and Justice. Cambridge, Cambridge University Press, 2000.

7 Chapman AR, Frankel MS (eds): Designing Our Descendants: The Promises and Perils of Genetic Modifications. Baltimore, Johns Hopkins, 2003.

8 Green RL: Babies by Design: The Ethics of Genetic Choice. New Haven, Yale University Press, 2007.

9 Habermas J: The Future of Human Nature. Cambridge, Polity, 2003.

10 Harris J: Enhancing Evolution: The Ethical Case for Making Better People. Princeton, Princeton University Press, 2007.

11 Resnik DB: Owning the Genome: A Moral Analysis of DNA Patenting. Albany, State University of New York Press, 2004.

12 Diogenes Laertius: Lives of Eminent Philosophers, Pyrrho. Book 9, pp 88–89.

13 Sextus Empiricus: Outlines of Pyrrhonism I. Book 15, pp 164–169.

14 Engelhardt HT Jr (ed): Global Bioethics: The Collapse of Consensus. Salem, M&M Scrivener, 2006.

15 Engelhardt HT Jr: Human nature genetically re-engineered: moral responsibilities to future generations; in Agius E, Busuttil S (eds): Germ-Line Intervention and Our Responsibilities to Future Generations. Dordrecht, Kluwer, 1998, pp 51–63.

16 Engelhardt HT Jr: The Foundations of Christian Bioethics. Salem, M&M Scrivener, 2000, pp 275–283.

17 Kuehn M: Kant: A Biography. Cambridge, Cambridge University Press, 2001, pp 391–392.

18 Kant I: Critique of Practical Reason. Beck LW (transl). Indianapolis, Bobbs-Merrill, 1956, p 126, AK V.122. G.E.M.

19 Anscombe GEM: Modern moral philosophy. Philosophy 1958;33:1–19.

20 Vovelle M: Revolution Against the Church. José A (transl). Columbus,Ohio State University Press, 1991.

21 Davies M: For Altar and Throne: The Rising in the Vendée. Forest Lake, Remnant Press, 1997.

22 Conn PM, Parker JV: Winners and losers in the animal-research war. Am Scient 2008;96:184–187.

23 Engelhardt HT Jr: Critical reflections on theology's handmaid. Philos Theol 2006;18.1:53–75.

24 Chrysostom J: Six Books on the Priesthood. Neville G (transl). Crestwood, St. Vladimir's Seminary Press, 1984, p 51.
25 John Cassian the Just Roman: Second conference of Abbot Joseph; in Schaff P, Wace H (eds): Nicene and Post-Nicene Fathers, Second Series. Peabody, Hendrickson, 1994, vol 11, p 465.
26 Ratzinger J, Habermas J: The Dialectics of Secularization. Schuller F (ed), McNeil B (transl). San Francisco, Ignatius Press, 2006, p 77.
27 Bradshaw D: Aristotle East and West. New York, Cambridge University Press, 2004, p 22.
28 Hierotheos [Vlachos]: The Mind of the Orthodox Church. Williams E (transl). Levadia, Birth of the Theotokos Monastery, 1998.
29 Hierotheos [Vlachos]: Orthodox Spirituality. Mavromichali E (transl). Levadia, Birth of the Theotokos Monastery, 1994.
30 Basil St: Ascetical Works. Wagner M (transl). Washington, Catholic University of America Press, 1962.

H. Tristram Engelhardt, Jr.
Department of Philosophy, MS-14, Rice University
PO Box 1892
Houston, TX 77251–1892 (USA)
Tel. +1 713 348 2491, Fax +1 713 660 7861, E-Mail htengelhardtjr@gmail.com

Pfleiderer G, Brahier G, Lindpaintner K (eds): GenEthics and Religion. Basel, Karger, 2010, pp 52–59

Eschewing Images of Man: Against Anthropological Reductionism in Bioethics

Friedrich Wilhelm Graf

Ludwig-Maximilians-Universität München, Evangelisch-Theologische Fakultät, Abteilung für Systematische Theologie, Munich, Germany

In his contribution to this volume, Tristram Engelhardt stresses several key points with which I essentially agree. It is true that in the interpretation of Kant, I would go further than he does. Kant was not only *not* an atheist, but indeed in his own particular way a devout Protestant. I find it difficult to follow some of what Engelhardt says about Eastern Orthodox theology. But I share, firstly, his strong emphasis on moral diversity, pluralism and radical religious diversity. In fact, in my estimation, the religious and moral diversity that forms part of our life-worlds needs to be given even greater emphasis than Engelhardt accords it. In the religious history of the modern period, specifically in the religious history of the 20th century, we see an incredibly dynamic process of differentiation. So far, our understanding of such processes remains highly inadequate. A great deal of what has happened in the religious life-worlds of the last hundred years cannot, or cannot yet, be adequately interpreted by scholars of religion. I will mention just one example to illustrate the dynamics of these differentiation processes. Around 1900, there were some 1,800 Christian faiths worldwide. Only 100 years later, there are at least 36,000 different Christian faiths. While there has always been a lot of divergence, fragmentation and schism in the history of Christianity – or, more precisely, of the various Christian faiths – religious history offers no examples of processes comparable to the modern and particularly the contemporary differentiation dynamics. The same is true of all those religious life-worlds that are known as 'Islam', 'Judaism', and so on. Collective terms such as 'Christianity', 'Judaism' and 'Islam' cannot be meaningfully applied. From the viewpoint of religious studies, they are indeed highly naive concepts.

Secondly, in my view, greater emphasis also needs to be given to perspectivism. Tristram Engelhardt claims that 'there are no moral answers from "nowhere"'. This

Translated from German by Jeff Acheson.

is also true in relation to religion: there is no neutral standpoint from which we can speak or think about religion. Here, our perceptions and thoughts are always from a given position or perspective. For we have each in our own way been shaped by a particular religious culture or by criticism of a religious or denominational culture. To this extent, it is only prudent to accept the contingency of one's own viewpoint. I myself come from a liberal-Protestant bourgeois life-world. Consequently, I can identify more closely with liberal forms of religion which promote the autonomy of the individual than with those forms of belief which in public religious discourse are known – somewhat stereotypically – as 'fundamentalism'.

I shall now explore these initial thoughts in more detail.

Religious Language: Radical Openness to Interpretation

What do we actually mean when we speak of religious bioethics or bioethics grounded in religion? The problems we face in the area of human gene technology are not to be found as such in classical religious traditions, texts or rites. The Psalms tell us nothing about the moral status of embryonic stem cells, and the Sermon on the Mount offers no insights into preimplantation genetic diagnosis. So what happens when venerable religious symbols and sacred texts of this kind are now applied to practical problems that have arisen as a result of advances in the life sciences over the past 40 years which are not only revolutionary and fascinating in equal measure, but also unsettling as they undermine traditional certainties? Ancient religious symbols, ritual images, narrative patterns and scriptural metaphors are taken up, reformulated and invested with new meaning by religious actors – Christian churchmen, Jewish communities, representatives of Muslim associations, etc. – and also by academic theologians in all kinds of ethics committees in an effort to address contemporary challenges. To this extent, it seems to me naive to speak of customs or traditions as sources of moral obligations in this context. More precisely, it is a question of reconfiguring a historical stock of symbols. This involves using old patterns of interpretation, with which a given actor wishes to achieve a present goal. This can be described as a method of 'suggesting' authority. An actor invokes certain religious traditions and then claims to be arguing 'on the basis of this tradition'. But this is not the case by any means. It is merely a constructive reinterpretation of traditions. This process also inevitably involves interests, discursive communication of claims to power and interpretative struggles. What is distinctive about religious language is that it is open to interpretation to an extremely high degree. Let me mention just one example, which is of great importance for discussions of bioethics. In current biopolitical debates, not only religious actors but also politicians and PR professionals in the media (journalists, etc.) are fond of invoking the Creation semantics which we find in many religious traditions. But Creation language is capable of numerous interpretations. For example, it can be used as the foundation of a universalist ethos, in which free individuals are endowed with dignity and equal rights. To quote the US Declaration of Independence of 4 July 1776, 'We hold these truths to be self-evident, that all men are created equal, that they are endowed by their Creator with certain unalienable Rights, that among these are Life, Liberty and the pursuit of Happiness'. This represents a truly classical political deployment of Creation language. However, the patterns of Creation theology can also be used to construct rigid ethnic religions, authoritarian communalization concepts and particularistic nationalist religions. This can be demonstrated with reference to numerous examples in modern Europe, notably in the 20th century.

If a high degree of openness to interpretation is a crucial structural feature of religious language, then in my view the bioethical or biopolitical questions which we debate need to be put differently. We need to ask about the authors of religious utterances and about the (situational) contexts. Where are religious ideas and concepts used, by what actors, and what claims to ethical interpretation or, more precisely, normative authority are being made in each case? Greater value should thus be placed on critical analysis of the use of religious language. For religious language – as theologians well know – is a dangerous language. Religious ideas are probably the most dangerous mental substance that we humans constantly have to deal with. Religious language is a language that can serve to transcend human limits. This transcendence is practiced by the devout when they use theological means to attain the same status as God. Anyone who, as a religious person, as a believer, manages to ascertain the will of God places himself, as it were, in an absolute position. He uses religious language to legitimize a position of superiority. This is a problematic use of religious language.

However, religious language can also be used in a quite different way, namely for the purpose of human self-limitation. We then speak of humility, of epistemological limits; we speak of that which we do not (yet) know. Numerous examples of both these ways of using religious language can be found in the present day. It should be noted that religious language is not only used by religious actors – in modern, open, globally interconnected worlds, the resources of religious language are essentially ubiquitous. Even efforts by doctrinal authorities to define certain interpretative boundaries for religious symbols are generally ineffective. Creation semantics can now be invoked by a wide variety of actors.

The Dignity of God Seems Not to Be Inviolable: The Abuse of Religious Language in Ethics

In biopolitical controversies, as can be observed both in European societies and in the US, people have no inhibitions about abusing traditional religious symbols as rallying concepts. Disputants use religious terms to dress up their individual, sectional viewpoint as a higher, unquestionable insight. Opponents of research involving totipotent stem cells cast themselves in the role of 'protectors of Creation', defending the limits of man as God's creature against modern-day genetic engineers. At the same time, a feature in a major German daily newspaper presented a chronicle of biotechnology under the heading 'Quest for creation' [1]. In the natural sciences, the unfortunate tendency to harness religious language can also be observed among scientists who carry out research in this field. These people hubristically no longer wish to leave man's future to God, but to usher in an 'Eighth Day of Creation' [2], to bring about Heaven on Earth. Scientists' interest in legitimizing the field in popular terms – which is itself a legitimate concern in the struggle for resources – encourages them to adopt religious language. In their own way, the churches also promote the inflationary devaluation of Christian symbolic language. Since the environmental debates of the late 1970s and 1980s, all the major churches in Europe have consistently abused the language of Creation in particular. Churchmen refer to the 'blessing of Creation', whatever this may be supposed to be. They speak of the 'community of Creation', 'responsibility for Creation' and 'fellow creatures'. Moreover, it must be seen as a sign of a political culture in crisis when religious language is incorporated into the Constitutional texts of ideologically neutral states. In my view, it is not a wise development to refer to the 'dignity of Creation' – as is unfortunately the case in the Swiss Constitution (Art. 120) – or to use religious terminology in drafting animal welfare legislation. Such inappropriate developments are the inevitable result of a failure to distinguish clearly between legal and religious language, or between the moral, the legal and the religious.

From the viewpoint of the history of religion and ethics, current bioethical controversies can be understood as a reenactment of ancient conflicts about law and the limits of human autonomy. We are not now arguing about anything essentially new. We are pursuing an argument which has been going on for 250 years, and which has been conducted notably by theologians. Nowhere in today's biopolitical discourse are generally original new lines of argument developed. All parties have recourse to concepts that have shaped ethical debate since the 18th century. Likewise, the use of religious symbols and theological ideas merely follows conventional patterns of use. The language of Creation is frequently used to brand modifications of nature as a 'Titanesque, sinful' abuse of human freedom. However, 'Creation' does not tell us anything about the consequences of our actions or possible unintended side effects, any more than 'Nature' prescribes particular norms for human action. Creation is an indispensable symbol in man's religious interpretation of himself and the world. But the term does not denote a source capable of yielding any normative knowledge of bioethical or biopolitical relevance. Reflective religious use of Creation language requires that man's ethical responsibility as a creature endowed with reason and free will should not be undermined by invoking some supposedly unequivocal certainties. No such unequivocal truths exist. Here, we must speak of hegemonic discourses and claims to power. Those who talk of Creation wish to invoke an authoritative source of normative knowledge which is not amenable to rational criticism in religious language games. It is clear from many texts in contemporary bioethical discourse that their authors are less concerned with the intrinsic meaning of religious concepts than with the strategic reinforcement of their normative decisions. In biopolitical debates, we are witnessing the inflationary use of religious terms, and inflation always entails devaluation.

Nowadays, in every conceivable kind of trivial dispute, reference is made to the concept of human dignity. This is a fascinating phenomenon. Theological encyclopedias of the 1920s, 1930s and 1940s do not include a single article on human dignity. Even in the 1950s, both the Roman Catholic Church and Lutheran churches took the view that human rights are an expression of a misguided liberalism, a modern mania for autonomy. Just 50 years later, these religious actors have secured a position wherein they now claim an unassailable monopoly in the interpretation of concepts such as human dignity and human rights. This can be described as the self-modernization of religious actors, who appropriate the language of modernity, while at the same time invoking time-honored ideas such as the traditional Judeo-Christian concept of man's being created in God's image, and claiming a unique interpretative authority over these traditions.

Another example can be used to illustrate the same problem. In current bioethical debates, reference is frequently made to a, or the, 'conception of man' [3]. In Germany, books are now being published on the conception of man in the German Constitution [4] or the Christian conception of man [5]. In earlier times, the socialist conception of man was a subject of intense debate within the Socialist Unity Party (SED) in East Germany. In present-day German politics, some claim that the Christian conception of man requires the prohibition of research involving embryonic stem cells, while others invoke their own conception of man in order, as it were, to distinguish different degrees of protection of dignity. But how can these *conceptions* generate norms? Do pictures of ourselves show us moral imperatives or prohibitions? Do free citizens under conditions of moral pluralism have to share a conception of man, or is it one of my basic rights in a free society to regard other people's conceptions of man as their own personal views? So long as questions like this remain unresolved, recourse to the 'conception of man' risks becoming an empty discursive public ritual. In general, it remains wholly unclear what the 'conception of man' is supposed to mean. Frequently, this term is used merely as a formula to suggest the existence of substantial agreement on solid foundations for coexistence within a diverse community, while no such fundamental consensus can in fact be determined.

Eschewing Images of Man: Against Anthropological Reductionism in Bioethics

In all these biopolitical debates, it is therefore important to reach agreement on terms and patterns of argument. One way in which such agreement may be sought is through conceptual history, the historical approach to concepts. Conceptual historians study the origins of terms, who has used them, and what interests are pursued by those who use them. The term 'conception of man', for example, has never been considered in the major philosophical or theological dictionaries of conceptual history, or in studies published by conceptual historians. We still do not know precisely where the term originated. There is a great deal of evidence to suggest that talk of the conception of man is of relatively recent origin, decisively influenced by discussions dating from the *Sattelzeit* – the period around 1800 marking the transition to modernity.

From the early 1950s – this is a peculiarly German debate – the 'socialist conception of man' played an important role in the East, and the 'Christian conception of man' in the West. Here, the 'conception of man' became a weapon in an ideologically charged civil war [3, p 139]. Moral theologians and social ethicists of the Roman Catholic Church subsequently linked this concept to modes of justification familiar from Stoic-Christian natural law. In line with the traditional idea of man created in God's image, they emphasized the creatureliness of man as the highest creature chosen by God, and Creation language now served primarily to set limits for man in relation to himself. In 18th century moral discourse, we can already find the idea that if man claims absolute freedom for himself, then he is sinfully and hubristically usurping the position of the Creator. In the debates of Catholic apologetics, autonomous man then becomes a kind of sinner par excellence. In current bioethical controversies, when church representatives warn against the 'Promethean' behavior of those who no longer respect natural limits, they are merely reprising the old warning against excessive human self-determination. While the 'conception of man' may be a particularly effective weapon in the biopolitical ideological struggle, it is not suitable as a fundamental concept for serious ethical reflection. For the phrase is notoriously vague. To use it is to imply that people could receive clear normative directions from a picture of themselves.

But the question arises how we are even to conceive of the origins and validity of the 'conception of man'? Two possible answers are to be distinguished. The conception of man is created by a non-human subject, such as God, from a specific external perspective – the way God sees man. Alternatively, we ourselves create images of ourselves through self-reflection in the mirrors of our self-examination. Conceptions of man are either products of perception by a third party or of holding up a mirror to ourselves. But the latter is devoid of ethical normativity, for why should the self-images created by self-reflection contain any orientational knowledge that is essentially new and not previously accessible? Either we have always known about limits set and norms prescribed for ourselves, or we lack such knowledge and are unable to attain it even in acts of reflective duplication through the production of images of ourselves. Moreover, the notion that we could gain previously unknown ethical insights through self-reflection remains dubious given the notorious fuzziness of the representation. How are we to know that we portray ourselves appropriately in our pictures of ourselves? Put differently, who guarantees the correspondence between the person producing the picture and the person represented? To this extent, speaking of conceptions of man involves numerous presuppositions and is highly problematic. Thus, the 'conception of man' is of little help in resolving current biopolitical problems.

The two main churches and their academic theologies have rich ethical traditions. Roman Catholicism, for example, has a fascinating stock of natural law theories, social doctrines and also theories of autonomous morality. In Protestantism – or in German Protestantism, at any rate – we find, since Melanchthon, the Aristotelian theory of the good; since Kant, concepts of the theonomous basis of autonomy; since Schleiermacher and Hegel, grand theories of functionally differentiated cultural spheres and an ethics of the objective spirit of institutions. But neither in Catholicism nor in Protestantism – and this is the weakness of these traditions – do we find much evidence of intelligent engagement with quite different conceptions of ethics, and especially utilitarian approaches, as practiced in Great Britain, in Scotland, and later also in the US. Even today, German theologians find it difficult to accept utilitarian forms of thought and patterns of argument. But within Europe, the ongoing process of economic, political and also legislative integration makes it increasingly important to reflect on the diversity of European ethical traditions.

A comment on this: what is interesting about the 'liberal' biopolitical positions recently adopted by the UK Parliament is certainly not what was adopted, but the debates conducted in the Continental European media. A recent article in the *Süddeutsche Zeitung* claimed that so-called Core Europe [6] – a notion that has been around for some years and seems first to have been used in 1994 as *Kerneuropa* by Wolfgang Schäuble and Karl Lamers – is represented by German idealism and sound German ethical tradition. Evidently, the First World War-era debates on the 'German spirit and Western Europe' [7] remain relevant for some ethicists 100 years later. At any rate, it is clear that we have great difficulty in engaging, especially in theological discourse, with certain other ethical traditions.

One fundamental problem with religious language is that it always seeks to be a language of completeness. It constantly speaks of the world as a whole, life in its entirety, the 'whole' man. From a religiously informed perspective, a person is indeed considered 'as a whole', but without making the quasi-totalitarian claim that anything like a definitive definition can be given of this whole. Religious language then serves precisely to qualify any sectional, monodisciplinary, positivist perspective on man. But then it also, as religious language, has the potential for self-qualification, and above all it implies a prohibition on seeking to put oneself in God's place. The person as a whole, in its individual limitedness or provisionality, can only be comprehended in life-historical terms, as a life history which is always a history of freedom. *Sub specie hominis* this whole always remains a fragment, and each individual can be said to represent essentially more and something other than that which it has made of itself or what we are capable of seeing therein. The most important insight of the traditional Judeo-Christian notion of man's being created in God's image is that none of us has a definitive definition of man. Man is to be defined as indefinable, and this seemingly paradoxical formulation symbolizes a fundamental individual freedom, which has always transcended any definition.

References

1 Kappes D, Makoschey K: Der Griff nach der Schöpfung: Die Chronik der Biotechnik: Von der Genforschung Mendels bis zum Klonen von Menschen. FAZ Feuilleton, 27 November 2001.

2 Judson HF: Eighth Day of Creation. Makers of the revolution in Biology. Cold Spring Harbor, Cold Spring Harbor Press, 1996.

3 Graf FW: Missbrauchte Götter. Zum Menschenbilderstreit in der Moderne. Munich, CH Beck, 2009.

4 Becker U: Das *Menschenbild des Grundgesetztes* in der Rechtssprechung des Bundesverfassungsgerichts. Berlin, Duncker & Humblot, 1996.

5 Huber W: Der christliche Glaube. Eine evangelische Orientierung. Gütersloh, Gütersloher Verlagshaus, 2008.

6 www.zeit.de/2003/52/Kerneuropa.

7 Troeltsch E: Deutscher Geist und Westeuropa. Gesammelte kulturphilosophische Aufsätze und Reden. Tübingen, Mohr, 1925.

Friedrich Wilhelm Graf
Ludwig-Maximilians-Universität München, Evangelisch-Theologische Fakultät
Abteilung für Systematische Theologie, Geschwister-Scholl-Platz 1
DE–80539 Munich (Germany)
Tel. +49 0 89 2180 3573, Fax +49 0 89 2180 2359, E-Mail ethik@evtheol.uni-muenchen.de

Pfleiderer G, Brahier G, Lindpaintner K (eds): GenEthics and Religion. Basel, Karger, 2010, pp 60–73

Children, Bodies, Life: Ethics as the Churches' Biopolitics

Petra Gehring

Institut für Philosophie, Technische Universität Darmstadt, Darmstadt, Germany

It is evident from the outset that questions of medical ethics are relevant to the canon of theological themes. Even bioethical debates extending beyond the medical sphere will touch on areas that can be described as 'theological', at least from the theological viewpoint: attitudes to creation (biodiversity, genetically engineered foods), to reproduction (artificial insemination, embryo research), to the body itself and its capabilities (enhancement), and questions of justice (access to biotech products, body-sale practices in the reproduction and human-tissue markets). Therefore, in this general context, the assertion that 'life' is a theological topic cannot be contradicted. Possible contextual references to these problematic issues, however, are not central to the matter at hand. This chapter is about activities and statements by theological players in the field of biopolitics. I am thus adopting what, from the viewpoint of theology, is an external perspective.

As examples – and also, I do not deny, in order to be provocative – I should like to address the churches' positions on three topics. Not primarily theological, they have been specially formulated for the bioethical discourse and are directly relevant to it. They are abortion, organ donation, and advanced healthcare directives – known as living wills.

I support the thesis that church biopolitics are not the only biopolitics (which nobody can seriously dispute), but that in the last three decades, taking abortion as its starting point, ethics (bioethics) has become a systematic instrument in public-relations work by the churches. Shortly after the war, what was known as the values judicature of the Federal Supreme Court [1] made rulings of the Supreme Court a governmental mechanism for the Federal Republic of Germany in fields including family and sexual policies. In the course of bioethics' career, however, church biopolitics has established itself to a previously unprecedented extent as the churches' own internally created culture of experts on bioethical issues, as an organized policy-advisory body and as a local-intervention agency: through confessional social work, or by engaging with citizens in bioethical debates with a strong public appeal. The field of life, according to my thesis, is by no means occupied solely by biotechnologies. It has also become the target for the direct expansion of theological lobbyism and the exercise of power by the churches.

Power of the Churches

Throughout its existence, the power of the churches has focused on the individual, on the question of spiritual guidance and on dealing with the body. The fact that the soul and the body are regarded as two different things is an extrapolation of the neo-platonic legacy, but also, of the history of power, a formula for success. In 'The Care of the Self' Foucault [2] argued that the duty of mistrust towards one's own body originated no later than the time of St. Augustine. This duty corresponds to two forms of domination that go very well together: the institutionalization of the instructional care of individuals, a supervisory 'pastoral power' [3], and the morality of prohibition, which invokes the perils of the sinful flesh and thus brings about the institutionalization of fear.

I do not propose to delve any deeper into past epochs of mistrust. Also the potent individualizing factor of moral self-examination, which we essentially owe to Christianity, would be a topic in itself. But let us get one thing straight. Even into the 19th and 20th centuries, within both Catholicism and the various strains of Protestantism, topics such as reproduction, abortion and married life were primarily characterized by a profound paternalism – wrapped up, of course, in the idea of God the Father and Roman law. Women and minor children belong to God, certainly, like all His creations, but in the eyes of family law they are first and foremost the chattels of the husband. They are 'entrusted to his care' within the context of a relationship of possession, which extends in the most concrete possible terms to the bodies of his wife and children. For almost two thousand years, the figure of the husband and father represented the fusion of the provider and good shepherd with the 'paterfamilias'.[1]

Against this backdrop, both matrimonial law and the law on abortion had a long history with countless twists and turns. As historian Günther Jerouschek succinctly puts it, the criminality of abortion – from the provisions of Roman law, through the mediaeval and modern periods to today – 'can only be thoroughly understood in the light of a conflict of interests resulting from the relationship between the sexes' [4, p 189]. The interests of women are not the issue here. It is primarily a matter of competition between men, the comprehensive rights of the husband in confrontation with the interests of a possible different father. Where abortion was prohibited, the discussion practically never touched on 'any personal rights of the fetus' [4, p 188]. Instead,

[1] Not until the 20th century did these relationships begin to loosen, for reasons that remain problematic to this day. Considerations of human rights, perhaps? Social hygiene, population policy? Wars?

arguments revolved around threats to the life and health of the potential mother – yet once again, what the law was designed to protect was not the woman herself, but rather the welfare of the family of the husband to whom she belonged. The church employed this sort of argument (at some times, though not at others) to restrict the absolute right of husbands to decide on the life or death of their families, limiting it to protecting the lives of their wives and the children born to them. But self-determination is not a factor here. At no time in the Roman law/Christian tradition was it 'left to the wife's discretion whether or not to abort' [4, p 189].

In certain respects, interestingly, the position of the churches in the modern age actually became more extreme at first. As the emergence of the welfare state gave rise to discussions about population policy, the churches increasingly came to accept even the death of the mother in exchange for the birth of the child. Women do not gain more liberty, but the child becomes important. This fits in with the fact that the criminalization of abortion focused increasingly on the pregnant woman: she was accused of wishing to avoid the burdens of childbirth and motherhood for base motives [4, p 195]. It was not until the vice squad met population policy at the end of the 18th century that abortion was added to the list of homicide offences, in which it had never previously figured in any legal system. The fetus still enjoyed no personal rights, though. Until well into the 1970s there was no concept of a 'developing' human 'being' in need of legal protection, nor was there any definition in the language of cell biology.

It was precisely this process – the establishment of a biologically based definition of the inception of life as the point at which the fusion of germ cells takes place, and thus at an extraordinarily early stage from the standpoint of legal history – that now constitutes a time of upheaval for the political self-positioning of both churches. Let me put it in provocative terms: the approach to biological definitions of life initiates the era of a new discursive involvement of the churches in the life sciences environment. Church biopolitics appears in its present form.

Human Life as a Biological Entity

The establishment of the developing human life as a biological entity requiring legal protection took place in stages. In 1968 the Vatican placed a powerful marker in the field of biomedicine with the encyclical *De Humanae Vitae*. Back then, like today, it was primarily the document's conservative postulates on reproduction policy that caught the public's attention. All medically based forms of contraception are forbidden, as are terminations and sterilization. The 'natural course' of 'conception', it argues, must not become a matter for human arbitrariness [5, no. 16]. Whoever 'adheres to the laws of conception does not behave as if he were the master of the sources of life, but rather the minister of the design established by the Creator. Just as man does not have unlimited dominion over his body in general, so also, and with more particular reason, he has no such dominion over his specifically sexual faculties, for these are concerned by their very nature with the generation of life, of which God is the source' [5, no. 13]. On the one hand the reference to the 'laws of conception' is intended as a legal prohibition, on the other the divine 'design' is mentioned, in which the 'generation of human life' has a permanent place. It is precisely in this connection that the word 'nature' receives its place.

Previous encyclicals had dealt with the inviolability of marriage.[2] *Humanae Vitae*, however, shifts the context of its reasoning and undertakes a revision of the concept of life in the theology of creation: the life that is being passed on is given biological qualities. The papal encyclical takes as its subject 'man's stupendous progress in the domination of the forces of nature, ... even of the laws that regulate the transmission of life' [5, no. 2]. And it calls for responsible parenthood to include an awareness of, and respect for, biological functions. 'In the procreative faculty the human mind discerns biological laws *(leges biologicas)* that apply to the human person' [5, no. 10], as the text has it – citing a passage from Thomas Aquinas where he speaks not of 'biological' laws but of 'natural law' *(lex naturalis)* [6]. The general substance of *Humanae vitae* is that biological knowledge apart, divine rules exist. But at the same time this means that 'the task of transmitting life' is subject to the prohibition of arbitrariness in harmony with nature and in recognition of biology.

In 1975, in a famous verdict, the Federal Constitutional Court struck down a provision approved by a political majority in the lower house of the West German parliament that unconditionally permitted abortion on demand during the first 3 months of pregnancy. At the same time the verdict went further than was required in terms of the case, finding that life begins at a point in time preceding nidation. The verdict thus picks up the spirit of *Humanae vitae*, at the same time creating a new subject of legal protection with constitutional ranking: 'developing human life'. Initially, however, nobody derived this protection under constitutional law directly from human dignity within the meaning of Article 1 of the Basic Law. Developing human life was thus not interpreted as a matter for absolute legal protection. Rather it was defined 'in the light of' the precept of dignity as a further relative good, and included as one of the many matters requiring legal protection listed in Article 2 (life, freedom from bodily harm).

The feminist controversy reached its peak in 1975. Was paternalistic 'protection' now to go beyond prohibiting actions 'on' a woman to procure her abortion, extending to matters within the female body? Does protecting the fertilized cell mean, in effect, that women are expropriated from themselves? Is a woman now to find part of her own body appearing against her as a legal entity in its own right? Interestingly, this new quality of access to the female body was quickly forgotten. The career of the unborn human life really took off in the 1980s, in the wake of the Greens' so-called 'Mothers' Manifesto' and thanks to a number of Christian protagonists in the

[2] See Leo XIII: Arcanum divinae Spientiae (1880) and Pius XI: Casti Connubii (1930).

environmental party and on the left. The philosopher Volker Gerhardt has rightly described its recognition, now virtually unanimous, as a phenomenon 'beyond all astonishment' [7]. German social democracy simply forgot about abortion on demand, and pro-woman GDR law was no cause for remembering it. It was equated to an indication solution. Critics of biotechnology and Christian bioethics became accustomed to the formulation that using an embryo meant killing it. And in 1993 the Constitutional Court, now an all-German institution, confirmed this once again. It established the principle of the womanless protection of life: the embryo, defined as an entity separate from the woman, must be regarded – not only vis-à-vis the pregnant woman, but also in all other respects – as in need of legal protection and state management. Discussion of embryo protection is now confined to its ambit. Its preconditions are no longer an issue.

One fascinating aspect of this process is the high acceptance level of the union of substantial thought in the field of creation theology on the one hand, and a concept of material life based on cell biology on the other. This supports a paternalistic, not to say patriarchal morality based on the legal-protection requirement, but which is on the whole compatible with the discourse on the 'value' of life, and also with the assessment of one good against another (against the woman, but also in favor of a medical indication). Compatible, at any rate, if you are prepared to forget the Kantian background to Article 1 of the Basic Law, which excludes any empirical reference for the postulate of dignity.

Another fascinating aspect is how the protection of the embryo brought Catholic pro-life campaigners and feminism together in the 1990s. Feminist bioethicists now cite the embryo as an autonomous human life – if not actually a legal entity [8].

And finally, the development of church bioethical opinion has brought Catholic and Protestant thinking together. Points of view are increasingly being publicized as jointly held, and the German Protestant Church has discovered the reference to 'life' and its 'sanctity' for itself. 'The protection of life' is the ecumenical program formula of the joint declaration 'God is a friend of life', published in 1989. Theologian Rainer Anselm has brilliantly shown that this approach to 'life' was alien to Protestant theory until well into the 1980s [9]. He shows that Protestantism actually has no tradition based on creation substance. Lutheranism, the dominant conservative culture, never adopted positions based on the protection of life: not in 1871 in the debate about the Imperial Criminal Code, not in 1909 when campaigners for women's rights called for the repeal of Article 218, and not at the end of the 1920s, shortly before National Socialism. Abortion was rejected on quite different grounds, citing familial responsibility and the purpose of perpetuating the race through the union of man and woman. The purpose of marriage is reproduction. Anselm [9, p 6] shows that even Dietrich Bonhoeffer strikes similar notes: 'In marriage man and woman become one in the sight of God, just as Christ becomes one with the Church'. 'God gives this union the blessing of fruitfulness, the creation of new life. Thus man participates in the will of the Creator through marriage. Through marriage men are brought into being for the glorification and the service of Jesus and for the increase of his kingdom' (Bonhoeffer as quoted by Anselm [9, p 6]).

After the war, the German Protestant Church continued to argue against abortion, but mainly on family-related grounds. And in the 1960s, when those opposed to the Catholic-inclined judicature of the Federal Supreme Court began to call more loudly for liberalization, demanding that the women affected should be free to make their own decisions on a matter of conscience, they found support among Protestant theologians [9, p 12].[3] The pregnant woman's freedom of conscience and her responsibility for family and reproduction are mutually antagonistic, but the same does not apply – as it does in the Catholic discourse – to the life of a woman (already a mother by virtue of having been impregnated) and the embryo's right to life.

But after the 1975 verdict of the Constitutional Court, the German Protestant Church veered in the direction of the protection of the embryo – and also, which is still more remarkable, towards a dogma of life that was initially expressed in the formula 'love of life'. No later than the declaration 'God is a friend of life' mentioned earlier, it arrived at a naturalistically underpinned position in which the ethically relevant duties of protection primarily apply to the unborn child rather than the woman. Conscience or dignity on the Kantian model are replaced by an 'unconditional right to life' and a dignity attributed to the entity in terms of biological substance – i.e. defined in terms of laboratory medicine – based on the concept of man having been created in God's image.[4] In place of the woman who has to reach a decision in a pregnancy conflict, the embryo has become the object of theological concern – and in the event that the embryo is not yet located in the laboratory, only its dependence on the woman 'who carries it within her' must be taken into account.[5] To a certain extent, then, what we have here is not a pregnant person but an embryo with a woman attached to it.

Bioethicization of the Churches' Positions

Must we speak of the 'bioethicization' of the churches' positions? I have a second example in mind: the field of organ transplantation. As far as I can see, the two churches

[3] Dealing with the post-war period, Anselm cites theologian Helmut Thielicke to show that the Protestant mainstream in the 1960s emphasized the moral conflict. Anselm generally interprets the Protestant position – at that time, in any case – as a discussion of 'the right attitude towards modernization' [7, p 2], not of 'life'.

[4] GBC/EKD 1989 [10]. The relevant sections are entitled Overview and VI. 2. The unborn life in the body of the mother.

[5] GBC/EDK 1989 [10]. The text expressly states that the embryo is 'not a part of the woman, but an independent separate being' [10, III. 4. The dignity of life before birth]. This separation from women gives rise to embryo-related maxims such as: 'anyone who wishes to protect embryos in the laboratory from research experimentation must also stand up for the life of the unborn child in the mother's body – and vice versa' [10, I. 3. The protection of life].

entered the discussion from the outset with a joint approach, at a time when, unlike the 1970s, bioethical debates were a permanent feature of the mass media.

The 1989 declaration 'God is a friend of life' described organ donation as acceptable in Christian terms[6], and in 1990 the German Bishops' Conference (GBC) and the Council of the German Evangelical Churches (EKD) published a joint declaration on organ transplantation [11]. This document reflects the existing social conflict about transplantation medicine only to a very limited extent, because the discussion focuses on the meaning of donation in terms of medical ethics and on dispelling concerns about what is known as brain death. What is particularly striking about the joint declaration is a passage that could have been written by an advertising copywriter, in which the medical details of organ transplants (including the especially contentious topic of donation by the living) are presented after the manner of a guide. They are declared to be both fundamentally desirable and practical. Here are a few samples at random.

Organ donation is described in the text as a gift, a gift of love. 'Life' is given a number of different shades of meaning. At one point the text speaks of the 'chance to donate one's life beyond death in love for one's neighbor' [11, Preface]. What is meant is an ethical sacrifice – and life in the non-biological sense. At another, however, the declaration speaks in much more technical terms of the donation of life, i.e. in the sense of the 'use' of life as the 'body' and corporeal substance. 'In the Christian understanding', it says, 'life and hence the body are the gift of the Creator. They do not belong to man to do with as he pleases, but, after carefully examining his conscience, he may employ them out of love for his neighbor'[11, 3.1.3 Ethical evaluation].

The paper expressly calls on the legislature to define the extent of the 'permissibility of organ donation by the living', which had not yet been done in 1990 [11, 2.2 Organ distribution]. It avoids from the outset the central problem of this contentious innovation from the viewpoint of medical law and ethics, namely the legal contradiction: since the operation to remove a donated organ is not conducted on a person requiring medical attention, the action of the surgeon is not designed to heal and the action of the donor constitutes injurious self-harm. Is the action of the surgeon at all appropriate for a doctor, does he not harm the donor, is he not exceeding the limits of his profession? The declaration touches only very briefly on this problem, reducing it to the question of whether the donor has given informed consent. Thus the language of the leaders of the political discourse at that time is adopted in the field of transplantation medicine. At the relevant point the medical injury inflicted is initially held to be 'justified and permitted by virtue of the consent of the patient concerned. But consent is only effective if the patient is rational and has had the nature and objective of the treatment (and, if appropriate, other treatment options) explained to him, together with the consequences,

[6] On a close reading, though, a certain caution is discernible: 'a Christian view of the human person (...) does not lead to the fundamental rejection of organ transplantation, but certainly to restrictive questioning. In general, the field of organ transplantation can be seen to be characterized by too much activity and too much experimentation' [10, VI.4. Organ transplantation].

dangers and risks of the procedure: if he knows what he is consenting to, in other words. As organ removal offers no medical benefits to the donor, consent is subject to special requirements' [11, 2.2 Organ distribution]. Thus the problem of non-beneficial surgery is unquestioningly absorbed by the reference to 'special requirements'.

Just like the concept of 'informed consent' borrowed from the legislation of English-speaking countries, organ donation – particularly by living donors – was the subject of fierce, critical discussion in German-speaking countries in the 1990s [12, 13]. One focus of the debate was the medical concept of brain death, introduced in the 1980s[7] as a legal precondition for the removal of organs from patients who were seriously injured or in a coma [14, 15]. Is the body of a dying man whose heart is still beating really dead? What counts? Indications of death that appear gradually, that are experienced by everyone, or cerebral criteria in the hands of a culture of experts?[8]

The churches' joint declaration leaves no room for doubt. Brain death, 'just like heart death', means that the person concerned is dead. A naturalistic definition of the person is provided: "The human spirit, unique among all living creatures, is physically linked exclusively to the brain' [11, 3.2.1 Definite ascertainment of death]. Public criticism of brain death is considered in the following sentence: 'The external difference between heart death and brain death may be mistakenly interpreted as if tissue and organs were removed before the death of the donor rather than not until after it'. The pursuit of commercial interests in the transfer of organs is condemned. But condemnation is restricted to a general statement that gives no further consideration to such questions as the origin of the organs allocated to recipients in Germany. Instead, the organ-allocation guidelines practiced by the semiprivate European agency Eurotransplant are praised as appropriate.

The paper ends with specific indications as to how the cause of organ donation can be advanced, i.e. how the subject of organ donation can be raised with the relatives of persons who are deceased or brain-dead: 'It may help the relatives if the doctor and perhaps the hospital chaplain tactfully enquire whether the deceased would wish to donate his organs to other people who are gravely ill' [11, 5. Concerns regarding relatives of deceased organ doners]. The matter of securing organs is thus linked with pastoral care.

Finally, it is remarkable – if we look at the opening and closing sections of the declaration – to note the solemn tones in which 'medical advances' are praised. Organ transplantation, it says, fulfills an 'ancient desire of mankind'. What once had to be accepted as the workings of fate has moved into the sphere of planning and responsibility, and this is raising new 'ethical' questions [11, 1. The state of affairs in tissue and organ transplantation].

[7] The Scientific Advisory Board of the German Medical Association formulated its first declaration on the diagnosis of 'brain death' (which goes back to the so-called Harvard protocol of 1968) in 1982, updating it in 1986, 1991 and 1997.

[8] There is a large volume of legal literature on the brain-death debate: see particularly Sengler and Schmidt [16]. For the history of brain death from the standpoint of the history of science and technology, see Schlich and Wiesemann [17].

The churches' joint declaration has rightly been perceived as an amazingly unambiguous adoption of a pro-transplantation position. To 'underpin' the determination of brain death, it makes concessions to practices in transplantation medicine and helps the doctor with his argumentation [18]. The involvement of representatives of lobbying organizations in the preparation of the joint declaration was also the subject of criticism [19], as was the fact that brain death was equated with death, which was seen in medical circles as implausible [20]. In the public debate on the advantages and disadvantages of transplantation, such an early disambiguation of the churches' position was not compelling – and nor was it universally welcomed at the church grassroots level. This made its effect as a public message all the greater. Particularly in the field of organ donation by the living, one of the paper's principal topics, the joint declaration of 1990 paved the way for a breach in the biolaw dam: the endorsement of the removal of whole organs from living healthy bodies as a resource for the politically desirable – not least on grounds of cost – the therapeutic strategy of organ transplantation. Eventually, following long discussions – and after a few a federal states had introduced their own legislation[9] – a national Transplantation Act was adopted in Germany in 1997. But brain death and organ donation by the living remained contentious.

Time Factor

My remarks on the third example focus still more closely on the time factor. Do the churches merely react to bioethical issues that are already under discussion, or do they take the initiative by raising subjects, guiding innovations in a particular direction, and thus creating biopolicy themselves? To put it more precisely: when during the last 20 years, and with what motives, have the churches entered into bioethical debates or the political arena? It seems to me that the answer must be: earlier and earlier, and with motives that are not at all clear.

A particularly good example is the proliferation of the advance healthcare directive, composed by the individual and expressing his wishes regarding the circumstances of his own death. First it must be stated that it was not introduced from the bottom up: the attempt to establish the living will, as it was initially called in the 1980s, did not originate within society – far from it. In the US the legal basis for living wills was created in 1976, and German specialists in medical law developed and announced an initial form for the purpose in 1978 [21]. From the very outset the subject was directly linked with assisted dying and transplantation. It was adopted by German organizations campaigning for the 'right to die'[10], and it also became an important

[9] Rheinland-Pfalz led the way with moral theologian Johannes Reiter, an advocate of transplantation who had been responsible for the earlier joint declaration, playing a central role.

[10] On the links between assisted dying and advance healthcare directives, sometimes involving the same individuals, see BioSkop [22] and (for more detail) Fittkau [23].

element in the establishment of transplantation medicine, on which the organizations that funded medical treatment were so keen. The mid-1990s saw the appearance of large numbers of possible forms designed ´to advance healthcare directives of one type or another. The German Humanist Association (a pro-assisted dying lobbyist with traditional links with the German Democratic Republic and a strong position in its former territory) got into the living wills business, and so did health insurers, clinics and solicitors. Some warned of 'high-tech medicine', promoting the form as an instrument of patient autonomy, while others, significantly more moderate, insisted that certainty had to be created and that caution was essential. In 1999 the German churches – once again acting in unison – put another living will on the market, this time called a 'Christian advance healthcare directive'. It was intensively promoted as a 'manual' for discussions between individuals and their advisers, and demand for it was brisk. 'Free from pain, fear and anxiety, and complying with the fundamental requirements of human existence, a middle way must be found for and with the dying individual between the unconscionable prolonging of life and its unjustifiable termination. In addition to medical treatment, human and spiritual care also have an important role to play' says the Press Office of the Evangelical Church [24], describing the purpose of the form.

Given the open nature of the 'Christian advance healthcare directive' in terms of both procedure and content, the question arises as to why it appeared at this time – or indeed at all. The text mixes statements about spiritual care with formulations that are awkward in the extreme: 'a life of human dignity', for example. But the main criticism of the document is that it uses extremely vague formulations, making it virtually meaningless in both legal and medical terms, and thus more or less useless to practitioners, except possibly as a way of initiating a discussion. This changed in 2003 with the appearance of a second edition with more legal substance. The EKD is now expressly calling for the advance healthcare directive to be given legal force.

The very fact that this Christian advance healthcare directive was at first so feeble seems to me to express in a nutshell the dilemma of the churches' current involvement in biopolitics. Does the bizarre model of a legally effective directive that is useless except as 'something to talk about' express certain reservations? Resistance to the juridification of how to deal with dying? Or are they just not entirely sure what 'Christian' contribution can be made to a discussion that is already under way? One way or the other, they want to be involved. The form of the directive is adopted and its biopolitical pattern amended (here the advance directive composed by the individual with sufficient legal and clinical clarity is the real joke of the form). Since bioethics is discussing advance healthcare directives, it is best to include a substantial dose of one's own ethics in a directive. Political questions such as who benefits from juridification, or whether the form of the directive is wanted at all, are thus avoided from the outset.

The churches are in any case anything but passive. They do not only react: they adopt the label 'ethics' in the English-language sense of applied ethics, with no ifs or buts. They want a public presence in debates, but they also want a role in giving direct policy advice – and in the last 20 years, from media work to the focused preparation of doctoral candidates, they have become highly professional. The Interdepartmental Centre for Ethics in the Sciences in Tübingen and the Institute of Science and Ethics in Bonn, established with considerable theological involvement in 1990 and 1993 respectively (and since supplemented and ennobled by the German Reference Centre for Ethics in the Life Sciences), are two prominent examples of personnel hubs in this field.

If the churches and theologians based in the science system adopt a proactive approach in the field of bioethics, as in the example, then naturally their positions with regard to content are adapted to the situation – as we have seen – in a relatively adventurous manner. In the case of unborn life, the Protestant Church breaks with both tradition and dogma incredibly quickly and for no discernible reason. In the case of organ transplantation, the two churches jointly press forward at a very early stage with an extreme pro-transplantation position, campaigning actively for the introduction of organ donation by the living. And in the case of advanced healthcare directives, they jump in with a document that initially, by its very nature, fails to do what it is supposed to do in technical legal terms – and without addressing any problems, either at the time or later. In the years that followed, the text was amended in line with legal requirements, and thus with the biopolitical context.

We shall now more closely consider which elements of the discourse come to the fore, and in what sense the churches' ethical input also relates to biopolicy – i.e. to policy designed to utilize and improve life substances, and thus to implement 'the careful management of bodies and the arithmetical planning of life' [25]. In the first and probably initial case, abortion and artificial insemination – amazingly, and certainly with 'ecumenical' side effects – a Catholic-style collaboration takes place between the churches to the detriment of the woman and the benefit of the unborn child. At the same time the switch to protection of the embryo is completed. In the second case, organ transplantation, imported text modules are used, for a good one third of the members of the commission of experts were specialists in medical transplantation and medical law, and also appropriately 'professionalized' theologians. The caste of the confessional bioethicist had made its appearance. In the third case, advanced healthcare directives, all it seems to be about is placing the statement function on the existing market in a 'Christian' form, then making sure the clinics have supplies 'at the end of life'.

Five Theses

'Ethics as the churches' biopolitics' – It would also be possible to go into personal interconnections, naming names in order to show that a group of ecclesiastically authorized figures, small though they may be in numbers, is behind the debate contributions cited above. Academic politeness requires that no notice be taken in a scientific context of what in other social contexts would be described as lobbying work, or of policy-advisory activities. I therefore prefer to remain abstract, concluding with five theses that I am happy to admit are provocative.

First: the churches are engaging in biopolitics. And they are engaging in it with conspicuous success and a new way, focused on the context of the fields that have been developing in Europe since the 1980: applied ethics, bioethics with its plethora of new forms of institutionalization in policy consultancy, mass-media PR, and also in the medical system as the preferred ethical adviser; in decision-making bodies such as ethics commissions; at the bedsides of the sick; in discussion with relatives, and at the side of the 'customer'.

Second: theological traditions are being pushed to one side in this context in both churches. Pragmatism is a part of bioethics in both the Catholic and Protestant discourse universes. Changes take place in accordance with the general climate in the politics of technology, with the policies of the day. The new pro-research position of Chairman Huber of the EKD on the Stem Cell Act might serve as a current example: the EKD is currently distancing itself from protection of the embryo, and against the backdrop of the pre-history explored above, we might well regard this as an adjustment to its previous shift – perhaps too abrupt – in the direction of the embryo.

Third: for the churches, the tendency in the 20th century away from the 'body policy' towards a 'life policy' congruent with the discourse of biology makes its appearance perhaps rather later than in debates conducted elsewhere, but today it is unmistakable. It would be ridiculous to depict the churches as the opponents of what Foucault called 'biopower' [26]. Perhaps, nonetheless, we should not necessarily analyze the 'bioethicization' – ugly word – of moral theology and Christian ethics in a theoretical sense solely as a phenomenon of biopower. Is the 'pastoral power' described by Foucault not also updating itself? Seen historically, does the self-empowerment of both churches to care for bodies and children *within* the woman, *within* the couple and *within* the family not have a very old aspect, perhaps going back as far as the Middle Ages? In any case the fact that family policy is conducted as population policy and policy on dying as policy on health are two long threads within the modern period that commenced before it. To this day the churches are not thinking about this, even though they explicitly wish to address realities in the medium of applied ethics. No more are they thinking about their internal paternalism, which today finds expression in their advisory activities with regard to abortion, transplantation and advance healthcare directives. And the guardianship of 'life' – not a positive factor in my view – that they exercise in those activities.

Fourth: it is unfortunate that a sociology of the newly appeared culture of experts on biological and medical ethics from 1970 until the present day has yet to be written. Why do theological and ecclesiastical figures get involved in the strange, partly scientific, partly publicistic, partly executive, discourse of bioethics? There is no doubt: it is at least partly for power and territory. Funding questions, the struggle for improved reputations, fights over seats on committees, the distribution of discourse power in policy advice: all these would be splendid subjects for one or more dissertations, so why not in theology too? Examining the importance of bioethical problems for the churches' public presence, we cannot avoid the impression that in recent years, bioethics has placed an incomparable public platform in their hands, together with an equally incomparable, if non-public, lever. Older ecclesiastical subjects – economic ethics, schooling and education, peace – are well behind biological matters in terms of their overall social importance, and hence well below them on the churches' agenda.

Fifth: bioethical subjects are also very obviously not only a way of reaching different publics, but also a way of reaching those affected by them. Is there a genuine 'demand' for bioethics on the part of the sick, perhaps? Personally I believe there is room for doubt. When we feel unwell, or anxious, or unable to make a decision, I think what we need most are friends – not adaptation to a discourse. But the pressure of public debates in any case increases the demand for expert knowledge, stimulating and maintaining the need for advice. It would be a subject in itself to investigate how, as well as active involvement in juridification, in moderation bodies that make democratic decisions and in the public coining of phrases, there has also been a change in the individual, pastoral dimension as the work of the churches has assumed biopolitical outlines. Possibly, however, the efficient route to the mobilization of residual religious feelings in the individual leads precisely to those fields on which biopolitics is concentrating today: to the beginning of life, to improving quality of life, to the end of life. Thus bioethics would be the modern mission in the theological sense, too.

References

1 Simon D (ed): Rechtswissenschaft in der Bonner Republik. Studien zur Wissenschaftsgeschichte der Jurisprudenz. Frankfurt am Main, Suhrkamp, 1994.

2 Foucault M: Sexualität und Wahrheit. Frankfurt am Main, Suhrkamp, 1986, p 167.

3 Foucault M: Sécurité, Territoire, Population. Cours au Collège de France, 1977–1978, Paris. Brede-Konersmann C, Schröder J (transl). Frankfurt am Main, Suhrkamp, 2004.

4 Jerouschek G: Lebensschutz und Lebensbeginn. Die Geschichte des Abtreibungsverbots (1988). Tübingen, Edition Diskord, 2002.

5 Paul VI: Humanae vitae encyclical, no. 16. 1968.

6 Aquinas T: Summa theologica (Summa theologiae). I–II, qu. 94, a. 2, 1265–1273.

7 Gerhardt V: Der Mensch wird geboren. Kleine Apologie der Humanität. Munich, Beck, 2001, p 14.

8 Gehring P: Feministischer Lebensschutz? Positionsverschiebungen im Vorfeld eines geplanten Fortpflanzungsmedizingesetzes. Femin Studien 2001;19: 83–92.

9 Anselm R: Abtreibung und Emanzipation des Lebens. Unpublished, 2008.

10 GBC/EKD: Gott ist ein Freund des Lebens. Herausforderungen und Aufgaben beim Schutz des Lebens. http://www.ekd.de/EKD-Texte/gottistfreund_1989_ueber.html.

11 GBC/EKD: Organtransplantationen. http://www.ekd.de/EKD-Texte/organtransplantation_1990.html.

12 Greinert R, Wuttke G: Organspende. Kritische Ansichten zur Transplantationsmedizin. Göttingen, Lamuv, 1993.

13 Grewel H: Recht auf Leben. Drängende Fragen christlicher Ethik. Göttingen, Vandenhoeck und Rupprecht, 1990.

14 Jonas H: Gehirntod und menschliche Organbank: Zur pragmatischen Umdefinierung des Todes; in: Technik, Medizin und Ethik. Zur Praxis des Prinzips Verantwortung. Frankfurt am Main, Suhrkamp, 1987, pp 219–241.

15 Hoff J, in der Schmitten J (eds): Wann ist der Mensch tot? Organverpflanzung und Hirntodkriterium. Reinbek bei Hamburg, Rohwohlt, 1995.
16 Sengler H, Schmidt A: Organentnahme bei Hirntoten als 'noch Lebenden'? Medizinrecht 1997;17: 241–247.
17 Schlich T, Wiesemann C (ed): Hirntod. Zur Kulturgeschichte der Todesfeststellung. Frankfurt am Main, Suhrkamp, 2001.
18 Doumat C Diagnostik des Hirntodes im Internationalen Vergleich; Diss. Universität Münster, 2005, p 33. http://miami.uni-muenster.de/servlets/DerivateServlet/Derivate-2675/diss_doumat/diss_doumat_constantin.pdf.
19 BioSkop 'Organspende' in Frage gestellt. Essen, Stattwerk, 1999, p 20.
20 Moskopp D: Zum Hirntodkonzept; in Moskopp D, Wassmann H (eds): Neurochirurgie. Handbuch für Weiterbildung und interdisziplinäres Nachschlagewerk. Wiesbaden, Schattauer, 2005, p 799.
21 Uhlenbruck W, Rollin M Sterbehilfe und Patiententestament. Berlin, Vahle, 1983.
22 BioSkop: Denkzettel no. 2: Patientenverfügungen. Essen, Stattwerk, 2001.
23 Fittkau L: Autonomie und Fremdtötung. Sterbehilfe als Sozialtechnologie. Frankfurt am Main, Mabuse, 2006.
24 EKD: Christliche Patientenverfügung. EKD und Bischofskonferenz veröffentlichen Handreichung. Press release, 1999. http://www.ekd.de/presse/697.html.
25 Foucault M: Sexualität und Wahrheit 1. Raulff U, Seitter W (transl). Frankfurt am Main, Suhrkamp, 1983, p 167.
26 Gehring P: Was ist Biomacht? Vom zweifelhaften Mehrwert des Lebens. Frankfurt am Main, Campus, 2006.

Petra Gehring
Institut für Philosphie, Technische Universität Darmstadt
DE–64283 Darmstadt (Germany)
Tel. +49 6151 16-4626, Fax +49 6151 16-3970, E-Mail gehring@phil.tu-darmstadt.de

Pfleiderer G, Brahier G, Lindpaintner K (eds): GenEthics and Religion. Basel, Karger, 2010, pp 74–81

On the Ethics Debate between Theologians, Scientists and Doctors: Experiences, Observations and Commentaries of a Medical Geneticist

Hansjakob Müller

Division of Medical Genetics UKBB/DBM, University of Basel, Basel, Switzerland

Premises

The amount of literature on the ethics debate between theologians and natural scientists is extensive. The purpose of this chapter is not to provide a survey of the papers previously published on this matter, but rather to describe the impressions of a medical geneticist who for many years has been striving for a constructive interdisciplinary dialogue on ethical issues, both within ethics committees and in the public domain. The main focus of the chapter is on the ethical problems that arise from the scientific progress made in modern genetics.

In today's world, religion and science are by no means totally separate from each other, although a certain distance has developed between theologians and scientists since the first scientific advances of the 16th and 17th centuries. Many theologians have, however, shown a distinctive interest in and knowledge of the natural sciences; the Augustinian monk Gregor Mendel (1822–1884), the founder of modern genetics, is a good example. However, with his valuable findings he might have exerted more influence on the contemporary debate on Charles Darwin's theories on evolution than he actually did [1]. It is a mute question whether his reluctance was due to his respect for the authority of the church or because of a lack of understanding.

There is no doubt that natural scientists emanate from the same 'world', that is to say: from the same cultural background, if not the same family, as theologians. Therefore, the challenges facing the two areas of expertise are not mutually unfamiliar. A scientist's daily work involves a profound and close study of creation, which permits him a very direct insight into its 'wonders' and helps him to develop great respect for its power. Most importantly, scientists are in a position to notice at an early stage of their research any questionable or inacceptable acts of interference that threaten the integrity of nature or the dignity of human beings. In a very practical way they are confronted with the ethical problems that result from advances in scientific research or from the implementation of its findings. Biomedicine is a case in point. Without doubt, scientists have a much more immediate knowledge of the burning ethical problems than theologians who – like other humanities scholars – sit in their 'cozy study chambers'. I remember very well how difficult it was in Switzerland in the 1970s to find theologians who were prepared to critically evaluate the ethical issues arising from the introduction of new procedures for medically assisted reproduction. Even if those days are gone, there is still a need to increase the intensity of debates between theologians and scientists in order to recognize and jointly evaluate emerging ethical problems at an early stage. This would lead to widely accepted conclusions with a sustainable impact.

Many ethical problems are arising in our time. The aim of synthetic biology is to generate biological systems that do not exist as such in nature. Man becomes a designer of microorganisms with new characteristics [2]. Human evolution has not reached the end of the road, either. The development of resistance to malaria in Africa or of milk tolerance in adults in pastoralist populations due to distinct mutations in the same DNA sequence that controls the lactase gene predominant in different parts of the world [3] may illustrate that evolution continues. Human beings even seem to be 'first-class evolvers'. Moreover, evolution is becoming increasingly complex. Genetic screening for a predisposition to diseases that have not yet manifested themselves, which already today is a heavily commercialized practice regardless of whether such testing makes sense or not, as well as perhaps germ-line gene therapy in the future, will contribute to this development. Linking up our bodies with robots or our minds with computers are other conceivable possibilities. So we must ask ourselves today: *Quo vadis, homo sapiens*? and, when do we raise a warning finger at *homo creator*, and for what reason?

Scientists can hardly conduct a fruitful discussion with theologians who have a fundamentalist mindset and, as a matter of principle, refuse to accept evolution and scientific perspectives that are not supported by the bible, which, after all, is not a textbook of science. It must be made clear to these theologians that moral principles, beliefs, and the feeling that our lives have a divine significance are not weakened but rather strengthened by scientific knowledge.

What, in the View of Doctors or Natural Scientists, Should Theologians Contribute to the Debate?

In the interdisciplinary ethics debate, as I understand it, theologians and other ethicists from the humanities are not necessarily required to provide academic reflections on terminology, principles or the essence of morality. However, they are expected to formulate comprehensible guidelines for complex but concrete ethical issues. With regard to solutions for concrete ethical problems, the guidelines described in the book 'The Principles of Bioethics' (first published in 1979) by Tom Beauchamp and James Childress [4] have won broad acceptance in daily medical practice. Procedural methods for specific situations can be defined and justified with the aid of four basic principles, although contradictions may occur regarding the interpretation of individual cases. The principles are: (1) autonomy, (2) beneficence, (3) non-maleficence, and (4) justice.

In medical genetics in particular, the counseling offered to patients often does not focus primarily on the scientific or medical aspects of a hereditary disease or a predisposition to a disease, but rather on its ethical or psychosocial implications. It is a question of providing effective support for fellow human beings in a difficult situation. Prenatal diagnostics, pre-implementation diagnostics, egg or embryo donations are the options to be discussed. Possible procedures for the future, such as germline gene therapy or justifications for the relevant research projects, are other related topics to be considered. Thus it is hardly surprising that medical geneticists turn to theologians or other professional ethicists for advice on how we may design our lives. Doctors who take care of people who are nearing the end of their lives are in a similar position. Like medical geneticists and specialists in reproductive medicine, they ask questions pertaining to applied ethics that can only be effectively addressed in an interdisciplinary debate.

My personal experience is: The more time theologians spend examining scientific and medical topics – particularly in ethics committees – the more likely they are to abandon their original positions and accept unexpected compromises. This occurs quite frequently. Some theologians even seem to enjoy demonstrating scientific facts – and they do so with didactic skill. In this way, they encourage the impression that these facts give little reason for concern. In the light of an increasing demand for bioethical evaluation, other theologians have quickly recognized a niche in the market, i.e. the lack of experts in theological ethics and, by showing a more pliable attitude, have found a comfortable tenure in the ethics committees of industrial concerns or political institutions. Such developments reduce the opportunity for a meaningful exchange of views. The churches and their adherents believe such representatives are not the right people to argue their case adequately. Many scientists, on the other hand, feel a need for a clear indication of the theological point of view and a statement of the attitude of the church.

What Meaning Do Natural Scientists and Doctors Attach to the Terms 'Facts, Data, Truth'?

Science and theology have basically different attitudes to their concepts. According to my understanding it is the aim of theologians to establish an ordered and enduring system of values. Scientists, on the other hand, use empirical data as a basis for new hypotheses and discoveries. Obviously, scientific paradigms are only constructs created by human beings and, therefore, short-lived and prone to revision. Opposition to them is by no means as emotional as discussions about theological viewpoints may be. However, this does not mean that the natural sciences do not defend their current 'truth' with an equally strong determination; naturally, they are also completely convinced of the validity of their scientific 'evidence'.

That being said, words do not suffice to fully comprehend and describe natural phenomena; their mechanisms cannot be squeezed into human thought patterns. The more scientists investigate and the more they appear to understand, the more surprises nature seems to have in store for them. It is no secret that every rule or discovery has its exceptions, at least in biology. Therefore, one of the essential duties of university teachers in genetics or related sciences is to remind their students that they must not be content to accept the current level of scientific knowledge, but must learn how to question apparent facts, i.e. the content of their curriculum, so that they can discover and build on new perspectives for the future. Examples of reversals in scientific thinking abound: In 1958, Nobel prize winner Francis Crick said: 'DNA makes RNA; RNA makes protein' [5]. This dogma of molecular biology, in its absolute form, is no longer acceptable today. The aim of the ENCODE (ENCyclopedia Of DNA Elements) project, for instance, which is the successor to the human genome project, is to identify and characterize all functional elements of the human genome and the transcriptome (all RNAs) [6]. Its findings are likely to threaten the validity of many other genetic principles that today's students have to memorize for their exams. Many such dogmata which are implemented in daily medical or genetic counseling practice today will also be subject to change. In accordance with Mendel's rules of hereditary transmission, it is considered to be a fact that a disease with X-gonosomal transmission, hemophilia A for example, can never pass from the father to the son, since the latter inherits the Y chromosome that lacks the gene for the blood-clotting factor. Nevertheless, there have been cases of sons who did inherit the disease this way [7]. They received both sex chromosomes from the father, the Y chromosome as well as the X chromosome with the gene responsible for the disease. The mother's X chromosome was lost during the early development of the embryo. This phenomenon is known as paternal uniparental disomy.

Certain theologians (and also other professional ethicists from the humanities) like to pounce upon such slightly damaged scientific dogmata, in order to either present their statements in a more interesting light or to throw doubt on the validity of all scientific facts in general. Neither attitude is very responsible, since such generalizations have no positive impact, but only cause insecurity among the general public.

Both theologians and lawyers have difficulty dealing with the unexpected vagaries of nature. Lawyers' linear way of thinking cannot fully comprehend the complexity of nature or of human beings. The legal profession would much prefer to have clearly defined stages in biological processes which it could use for the formulation of laws, as becomes apparent in the debate on the beginning of life or the definition of totipotency. Life does not start at a fixed point in time! It already exists in the germ cells. Even after the combination of the germ cells in the zygote, which is a complex process, the development of several embryos is still possible. Several individual embryos may emerge from the zygote up to 14 days after fertilization. The concept of the unique totipotency of the early embryonic cell, which was put forward as a strong argument against extracting such cells for diagnostic purposes in pre-implantation diagnostics, has been seriously questioned by recent results in stem cell research [8]. Moreover, it is unlikely that a single cell taken by biopsy from an early embryo can ever develop into an infant since it lacks the structures that are essential for implantation into the mucous membrane of the uterus, the zona pellucida.

Beware of Simplification, Generalization and Polarization

What is 'theology', what is 'science'? The natural sciences encompass many different fields of study. Their spectrum ranges from mathematics and the so-called 'exact' sciences, such as physics and chemistry, to geology, paleontology and modern biology. The representatives of the various disciplines differ as to their interests, their scientific background and also as to their intellectual abilities with regard to abstract thinking or visual perception. Views can differ greatly even within a single discipline. Often this fact is not sufficiently taken into account. People often talk about the 'opposition' and simplify the issue by creating two camps, the 'scientists' versus the 'theologians'. Theologians of the same church may have different career paths, different abilities or temperaments. For example, a catholic theology professor, one of my colleagues in the Swiss delegation to a previous bioethics commission of the European Council, always had a sharp word to say whenever the representative of the Vatican, a mild-mannered bishop, expressed his opinion on the ethical evaluation of biomedical measures.

For participants in interdisciplinary dialogues on ethics it is difficult to estimate the professional competences of members of other scientific communities. I have observed, e.g., that some theologians like to boast about their contacts with well-known and leading representatives of the genetics community who are engaged in top-level research and with whom they have debated ethical issues. However, they fail to realize that the persons in question come from the exact sciences or are specialists in information technology and thus have but few connections with biology or the wide scope of the genetics profession. This only brings one-sided views on the facts and on scientific progress into the ethical debate.

Particularly in the German-speaking sphere, genetic research is treated with a great deal of suspicion, which often has a negative effect on the implementation of its findings in daily medical practice. There are unfounded fears relating to reasonable suggestions on how to deal with problems of a genetic origin. This state of affairs must not be made more difficult by subsuming all the ethical problems in modern biology and biomedicine under the heading of genetics. Even the abbreviation, GenEthics, however attractive it may sound, should be avoided. We also cannot speak of 'NeuroEthics' or 'GerontoEthics' as separate disciplines, although there are undoubtedly particular ethical problems that arise during the development or the implementation of procedures in neurology or geriatric medicine.

It is also the media which can influence the debate in that they tend to give more access to natural scientists with a so-called atheistic perspective, although as a rule they only represent a minority view. The almost 'sectarian' manner in which they loudly broadcast their opinions betrays the lack of foundation in their arguments. In this regard there is not much difference between them and religious fundamentalists.

Who May Give the Answers?

In medical genetics, paternalism is a thing of the past. Patients seeking advice should receive the fullest possible information, but they have to decide for themselves how to proceed. In the same way, all holders of 'moral beliefs' should be given the freedom to express their opinions freely; they are part of our society, which wants to or should have its own system of moral values. The autonomy of the individuals participating in the debate must be guaranteed. For churches with hierarchical structures and fixed dogmas this requirement is more difficult to meet.

Increasing mobility means that doctors come into contact with more and more people from other cultures and other religions. Therefore, it is important to learn more about other belief systems and their moral principles. In my opinion, this does not mean, however, that the theological colleges are required to carry out a theoretical revision of the rules of morality.

Fostering an Interdisciplinary Dialogue

Ethics and religions cover common ground. We must learn to understand our history and the myths that govern our daily lives. The natural sciences, on the other hand, constitute an inexhaustible source of knowledge of the world and of the human beings that inhabit it. This knowledge must be made use of since in a certain sense, this knowhow also has a normative power. Attempts to bring scientific findings under the control of the church or other religious institutions will inevitably lead to a dead end.

The interdisciplinary dialogue between workers in natural sciences and biomedicine on the one hand, and academic theologians and/or proponents from religious institutions on the other, must be intensified. Clever observations of individuals who try to reconcile scientific findings with aspects of their own beliefs are not sufficient. The debate must involve a broad section of the fields of expertise concerned in order to meet with an adequate response in the general public. Therefore, the churches should offer more opportunities and organize more events to address the specific ethical questions that we have to deal with in our everyday lives.

Communication between theologians and natural scientists or doctors has been made more difficult now that fewer and fewer scientists and doctors have a classical high school education including Latin and Greek. Therefore, many of the terms used by theologians are unfamiliar to them. On the other hand, the language of today's natural scientists with all its abbreviations and catch phrases must be difficult to appreciate for a theologian. Interdisciplinary discussion therefore requires practice. It is important to develop an understanding of each other person's 'language' and manner of communication. Because of the multitude of data generated everyday by scientific and medical research, the transmission of information must be as concise and specific as possible and make use of audiovisual aids. The representatives of these fields are no longer accustomed to listening to long, purely oral lectures. Here, the universities could help by organizing more interdisciplinary seminars on current ethical issues.

The interdisciplinary ethics debate is a challenge to us all! Research in the natural sciences is advancing rapidly, last but not least thanks to the opportunities offered by bioinformatics. For the individual it becomes increasingly difficult to follow all the separate developments and place them in an ethical context. Only an open dialogue between theologians and scientists will enable us to more or less keep pace with scientific developments and establish and follow a modern, up-to-date system of ethical values.

Discussions between theologians and natural scientists will always contain consonances and dissonances, but differences of opinion may constitute a valuable source of inspiration helping us to find our bearings in a complex world. Differences of opinion do not mean that dialogue cannot be held in an open and constructive manner. The understanding of and the respect for the other person's opinion is an essential requirement for a fruitful discussion.

References

1 Klare R: Gregor Mendel. Father of Genetics. Springfield, Enslow, 1997.
2 Endy D: Foundations for engineering biology. Nature 2005;438:449–453.
3 Enattah NS, Jensen TGK, Nielsen M, et al: Independent introduction of two lactase-persistence alleles into human populations reflects different history of adaption to milk culture. Am J Hum Genet 2008;82:57–72.
4 Beauchamp TL, Childress JE: Principles of Biomedical Ethics, ed 5. New York Oxford University Press, 2001.
5 Crick FH: On protein synthesis. Symp Soc Exp Biol 1958;12:138–163.
6 ENCODE Project Consortium: The ENCODE (ENCyclopedia Of DNA Elements) Project. Science 2004;306:636–640.

7 Vidaud D, Vidaud M, Plassa F, Gazengel C, Noel B, Goossens B: Father-to-son transmission of hemophilia A due to uniparental disomy. Am J Hum Genet 1989;45(suppl):889.
8 Conti M, Giudice L: From stem cells to germ cells and back again. Nat Med 2008;14:1188–1190.
9 Ahn AC, Tewari M, Poon CS, Philips RS: The limits of reductionism in medicine. Could systems biology offer an alternative? PLOS Med 2006;3:e208.
10 Arber W: Grundlagen der Gentechnolgie; in Müller Hj (ed): Reproduktionsmedizin und Gentechnologie, Basel, Schwabe, 1987, pp 138–153.
11 Schuster S, Eils R, Prank K: 5th International Conference on Systems Biology. Biosystems 2006;83: 71–74.

Hansjakob Müller
Division of Medical Genetics UKBB/DBM, University of Basel
Römergasse 8
CH–4005 Basel (Switzerland)
Tel. +41 61 685 64 33/32, Fax +41 61 685 60 11, E-Mail hansjakob.mueller@unibas.ch

Pfleiderer G, Brahier G, Lindpaintner K (eds): GenEthics and Religion. Basel, Karger, 2010, pp 82–100

'Biopower' – Ethical and Theological Considerations

Dietmar Mieth

Max Weber Institute for Advanced Studies, University of Erfurt, Erfurt, Germany

Biopower: Moratorium Instead of Ban?

Biopower, in my view, is the connection between Bioscience, Biotechniques and Economy. In general, the connection between Science, Technique and Economy may be legitimate because the development of modern societies is dependent on a kind of 'covenant' with this connection. On the other hand, the connection in the field of biotechnologies can also work as a pressure group in the lobbyism of political decisions. Biopolitics then may depend on the power of this connection. Even in advisory groups on ethics, it is self-understanding that there are scientists who must judge projects that they want to promote. In other advisory committees, economy – in this case pharmaceutical interest – may well be represented. The independence of political decisions may be juridically guaranteed but cannot be ensured.

On the other hand, ethics is not requested because power is a priori suspicious. But power in democracies must be controlled by public discussions and institutions. In my opinion one of the signs that bioscience is powerfully connected is the use of a 'moratorium', when there are fundamental objections against the implementation of a new technique in the area of reproductive and genetic engineering. A moratorium gives time for reflection and therefore may be useful if we do not know all consequences of the implementation. On the other hand, principles like human dignity, if they are taken as a criterion to evaluate all other aspects, are not touched by the questions of (non)utility. A moratorium like this, searching for time to discuss and to decide, is ambiguous. It also gives time to the power of economical interests. Economical interests are connected to a utilitarian view. Nevertheless, the utilitarian view of ethics – which I do not share – does have some advantages. The following example shows this problem.

On the occasion of the public announcement of the Dolly Cloning in the journal *Nature* (New Scientist 1997), scientists, politicians, and ethicists emphasized the broad consensus that cloning human beings is not to be allowed and that such a perspective should be ruled out by national and international bans and conventions. A certain, though not fully discussed, recourse seemed to be the Human Rights Convention on Biomedicine of the Council of the European Union with its prohibition of the intentional transmission of genetically manipulated genotypes to the next generation. On the other hand, in the journal *Nature* (New Scientist 1997) the thesis was immediately advanced that 'human cloning' would require a 'moratorium', but not a 'ban'. The fronts are clear, and they proceed similar to those in the debate on so-called germ-line cell therapy, the gene transfer in the sex cells for the purposes of genetic modification of future generations. Here too, the one side (scientists, but also ethicists) assert that if the method can be used precisely, safely, and without side effects, with the goal of eradicating monogenetic hereditary diseases, one is morally obligated to make use of it to spare entire generations immeasurable suffering. The other side wants to do without every manipulation of this kind, above all because of the uncertainty of its consequences for a still unknown future context, and also because one cannot consult those who will themselves be affected in the future on such a serious matter, and finally, because the excessive use of embryos in experimentation has been rejected as the unavoidable way to germ-line transfer up to now. In light of this controversy, the EU Advisory Group on 'Ethical Implications of Biotechnology', in its position paper on gene therapy (1994), cautiously formulated that so-called germ-line therapy cannot be allowed, on technical as well as ethical grounds, 'at the present time'. Is this 'present time' still ongoing?

Anyway, it can be stated that the ethical discussion on the cloning of human beings is not limited to arguments demanding a strict prohibition on principle. With new technical or technological possibilities a previous 'taboo' often disappears; by 'taboo' I mean a prohibition which has not been analyzed and which is seldom under pressure to be justified. Examples are in vitro fertilization (IVF) with embryo transfer and prenatal diagnosis (PND). Here the loss of embryos and fetuses has become routine. The methods are selective, and continue to be ethically problematic, even though they are legally permitted.

On closer scrutiny there are three positions on the question of human cloning: an abstract endorsement under almost unattainable conditions, so that this actually constitutes an option for a moratorium; a rejection that wishes to defend our world of everyday life against radical interventions, and an equally strict rejection that is justified not only by categorical but also by a cluster of relative arguments.

The conditions for the abstract endorsement: high-ranking goals compatible with human dignity; precise and safe implementation; no negative effects on contexts and future worlds. These conditions seem to me abstractly desirable but not concretely attainable. Another objection is that the goal of replacing diseased or defective organs with others from a warehouse of human spare parts is just as incompatible with human dignity as the goal of replacing a dead child with an exact copy. The principle of human dignity as formulated by Immanuel Kant says that the human being is never to be treated only as a means but always as an end in itself as well. This state of being an end in itself would be lost in acts of procreation for purposes of instrumentalization. A desire for another child is understandable, a child copied on request is by definition a replacement. Equally problematic is the goal of permanently controlling genetic diseases, not through the germ line but through cloning. The uncertainties about its consequences for changing environments as well as the absence of possibilities for acquiring the consent of the human beings affected by this new world of everyday life are the crucial arguments here. One suspects that the position of a conditional endorsement under negotiation has been introduced for strategic purposes: if one agrees to it, the next question is automatically and immediately what degree of risks one is willing to accept concretely after having already agreed in abstracto. I consider this position less an attempt to justify this ethically than an attempt to outwit the inexperienced participants in the discussion. Or to use a metaphor appropriate to the context 'Dolly', it is an attempt to pull the wool over the other participants' eyes. This is nothing new: first present or coming problems are excluded from the question of permissibility, then once a basic consent has been attained, the dossier is reopened, and one struggles more or less futilely with the problems. In my opinion, the principle valid here is: one should not solve problems in such a way that the problems that arise through the resolution of the problem are greater than the problems that are being solved.

With human cloning, scenarios of potential worlds of everyday life become visible, scenarios that are truly shocking, especially if we consider our morality a product of our everyday world. This world of everyday life is the world of our daily experiences, of our becoming individuals in a social context, our societal life as a communicative process. Science and technology are part of it, but they must remain incorporated in this world of everyday life. When 'creation' or 'nature' are mentioned by religious individuals or by those with 'religion as heritage' (Ernst Bloch's 'Religion im Erbe') in this context, these symbolic words address a stable point of reference in this world of everyday life. We could also speak of the achievements of human history and culture. These include our culture of sexuality and the situation of human reproduction in identifiable relationships. The vision that sexuality, because totally separated from reproduction, will one day be reduced to repeatable brain stimulations is just as great a culture shock as the vision that human beings will one day intentionally be assembled from different, 'biological' materials.

The human being reacts to the shock threatening our world of everyday life with taboos. Taboos offer a sense of protection, and act as a wall against everything that appears to threaten the goods vital to the world of everyday life. 'Taboo' has, of course, become an ambiguous word, sometimes even an openly abusive word, but the rationality of the apparently irrational conceptions (irrational because not spontaneously justifiable) can often be shown when challenged. Hans Jonas's 'hermeneutics of fear' ('Hermeneutik der Furcht') – concerning options for future worlds one should sooner pay attention to that which is to be feared than that which is to be hoped for – is, in my opinion, nothing other than the proof that behind taboos negative experiences are to be found, for example, experiences like those with the selection of human beings in the 'Third Reich'.

The second position, the rejection of cloning, wants to protect the goods of our world of everyday life from change, a position upholding conservative values. Spontaneous statements by politicians frequently reflect this consensus in conservative values. But how long will it hold? How long will our world of everyday life, which is constantly changing, remain a stable source of resistance? Won't politicians find it easier to limit the most drastic outgrowths of misuse, on the one hand, and allow each individual to make use of technological progress according to his or her options on the other? The individual option is always the option of decisive individuals, to which developing or future human life will have to conform. In light of such prospects, all too well confirmed by the experiences of the last decades, we must attempt to come up with a justification for the rejection of human cloning which is not only based on one single categorical argument but instead resembles a cable in which several arguments, that are possibly relative if taken alone, are bound together and can act as resistance together.

This is the third position, which I will argue for in the following: the clustering of relative arguments that act together as a strict prohibition. The *first*, relative argument arises from the weaknesses of the first conditionally permissive position which I have already presented. A *second* argument is described as the non-instrumentalization argument. According to this argument, each human being, even if he or she does not have the right to his or her own unique identity, has the right not to be constructed as a copy of a second human being in accordance with the planning of a third person. With cloning it is, of course, conceivable (depending on the '[surrogate] mother animal' bringing the fetus to term) that no perfect copy, that is to say no copy of a total identity, is produced, because the adaptation of the cells can differ, and because the environments can also have a modifying effect. Nonetheless, I still think that in our acting as responsible human beings – 'nature' bears no responsibility, for example, for the creation of identical twins – we must orient ourselves towards the highest achievable possibility in unique identity. Even if 'Identity' is thus not a categorical against human cloning, the non-instrumentalization argument is categorical in most of the cases which we can imagine.

Third, the road to 'Dolly' was literally paved with fetal experiments. We must confront ourselves with the moral question – which already applies on the level of mammals, although not categorically – whether we can take the responsibility for experimentation entailing such an excessive number of fetuses, or in general for a method entailing trial and error to such a large degree. The fourth argument is analogous to the prohibition of germ-line transfer (on the basis of uncertainties that may arise and the absence of the 'informed consent' of the human being affected under changed circumstances). Here it must be added, as with germ-line transfer, that the borderline between illness and 'design according to desire' has become indistinct. If illness (in the definition of the World Health Organization on health) is the limitation of physical, psychic, and social wellbeing, then the borderline between therapy and the genetic improvement of the human being can quickly be crossed subjectively. This is – the fifth argument, the argument with the slippery slope, where every application lands, no matter how limited initially. Who will still be able to determine the individual borderlines, when the borderline as a whole has been given up? It is also a relative ethical argument to weigh the alternatives against each other, to decide which line is best drawn and maintained.

A sixth argument takes up the uncertainty concerning our world of everyday life. Can one support the family as the 'nucleus' of society and, at the same time, risk its structures, even if apparently marginally for the time being? The conception that women will one day donate the eggs, and men will warm the incubators with the cloned, designed children under their skirts may seem satirical and abstruse but perhaps the day will come when someone comes up with the idea that this is not without precedents in nature, for example, the male albatross caring for the brood. Are there certain things or goods that we do not wish to give up? Are there limitations and pain and suffering that we are willing to retain because the price for their elimination is too great? These are serious questions, and it is our moral responsibility to find answers for them.

These six arguments act, to return to my metaphor, like a cable constructed out of six different strands. Individually, the arguments may be relative, except the non-instrumentalization argument, in most of the thinkable cases. But the individual arguments cannot totally dismantle every thinkable counterargument. Nevertheless they still keep their argumentative plus and thus strengthen the position of the ban on cloning. A wall built of individual stones can be as sturdy as a boulder. Yet the individual arguments are still open to discussion under changed assumptions, contexts, and consequences. The technical possibility can and will challenge ethical reflection but it cannot replace it. As long as we understand our culture as the sum of our societal life expressions, our world of everyday life cannot be and cannot be allowed to be determined one-sidedly by an alliance between science, technology, and business. Therefore: not only a moratorium but a ban on human cloning.

In vivo Cloning and in vitro Cloning – Biopower in the 'Politics of Language'

Our language is not capable of grasping all new phenomena immediately, thus any concept one can choose has the potential of being misunderstood and this may also be applied to the differentiation between 'reproductive' and 'non-reproductive' cloning. 'Reproductive' means that the clone is implanted and a human adult person develops from it. In vitro does not yet determine whether an early embryo, which has been manipulated accordingly, will or will not be implanted. And if it is not implanted, but rather experiments are carried out on it with particular long-term therapeutic goals, then this counts as 'non-reproductive' cloning.

The ethical advisory group of the EU came to the conclusion that 'reproductive' cloning must be forbidden. The Clinton Commission, using the same terminology, was more liberal only insofar as it demanded a moratorium 'at present'. The formula 'at the present time and the present social context' did in this case not enter into the conclusion of the European advisory group. In my opinion, this limitation assumes that this kind of prohibition will intend its own revision! It is self-evident that every rule which we formulate under the present circumstances can be revised in the future. Yet whoever wants to bring this to expression has ulterior motives which have to be put on the table. Thus it can be said that the Clinton Commission declared itself in favor of a moratorium whereas the European advisory group was in favor of a strict ban, but only in the so-called 'reproductive' area. (I will come to the reasons behind the strict ban later.) In the 'non-reproductive' area, that is, in vitro cultivation of cloned embryos in cell cultures, the EU advisory group stated that in those countries which allow experimentation on embryos (such as, Belgium or Great Britain) in vitro cloning should not be forbidden as well on condition that one is dealing with high-priority therapeutic purposes, as well as on condition that a licensing body, that is, an ethics commission, is consulted, and finally, on condition that the manipulated embryos will not be implanted and become independent human beings.

In my opinion the differentiation of 'reproductive' and 'non-reproductive' cloning was not decided on a factual basis; it rather intended to reach a differentiation in the treatment of 'embryos' and 'humans'. In a 1994 document I found a statement by the ministerial council of the European Council regarding tissue banks in which 'reproductive' and 'non-reproductive' were defined as to how they are to be understood in this arena. There, egg cells, sperm cells and embryos were considered to be 'reproductive'. Their use as tissue was to be forbidden exactly because they are 'reproductive'; that is, from these tissues human beings could develop. The expression 'reproductive' thus was intentionally narrowed down to implanted embryos by the advisory group in the interest of practicability. This I call the politics of language. This politics was not without precedence. This difference between 'reproductive' and 'non-reproductive' has previously been introduced by an American advisory group on IVF.

Non-reproductive cloning today is called 'therapeutic cloning', or otherwise, this is also 'biopower' in the meaning of the politics of language. The concept of therapy or the concept of human health plays a very central role in the politics of language, therefore I would like to elucidate this with an example from the Convention on Biomedicine of the European Council. Article 18.2 of the convention states that the production 'of embryos for research purposes' is forbidden. Article 12 implicitly deals with research on embryos, although explicitly it is concerned with the possibility of sex selection. There it is stated that this is possible only if it serves health purposes. As an example of health purposes, sex selection in order to prevent hereditary disease is mentioned since a hereditary disease can be connected to a particular gender. The text does not clarify about which method it is speaking: abortion after PND, embryo selection before implantation, possibly also sex selection through semen centrifuging.

At the same time, Article 12 states that these health purposes also include 'research on health purposes'. Of course, one wonders at this point what is meant by the prohibition of embryo production for research purposes in article 18.2? Are research purposes for health purposes, which elsewhere are always allowed, excluded? If this were the case, then the research purposes of article 18.2 would mean little, because one could declare any research purpose in this area to have a therapeutic purpose.

The Biomedical Convention as well as its supplementary report on cloning which was signed in January 1998 leaves the establishment of what is a human being to the nation-states. In an interview in the German weekly journal *Focus*, the president of our advisory group, Noelle Lenoir, explained that when human life begins is relative to the nation and its respective laws [1]. One can imagine that, in the attempt to reach a valid European consensus, the openness of this question results in any consensus remaining unclear in the specific case. According to Lenoir, Germans consider an early embryo to be a human being, other countries only recognize individual human beings after birth. In the Human Rights Convention on Biomedicine it is still not determined what a 'human being' is.

This has particular significance for the cloning of human beings. In the supplementary report (January 1998) it is stated that it is forbidden to create a human being with the intention of making him or her identical to an already living or deceased person. Someone reading this without bias assumes that this means that any human cloning is forbidden. Yet things are not so simple, even if in the second paragraph it is stated that this prohibition is so strict that there are no exceptions. However when consulting the official explanation, things look different.

In this explanation it is stated that one has to distinguish between three levels of cloning. First, the cloning of cells in general is morally not problematic. I am also of this opinion. Cloning is not morally problematic when dealing with a living being without an independent destiny. Second, in vitro cloning, here 'embryo cells' are carefully looked at. But behind 'embryo cells' embryos in the totipotential state are hidden. Third, the cloning of 'human beings'. This means that the cloning of 'human beings' is being distinguished from the cloning of embryo cells, and then it is expressly indicated that this supplementary report applies only to the cloning of 'human beings'. Thus, I come back to my discussion of the politics of language: although the expression 'non-reproductive cloning' is not used here, it is clear that the technical differentiation which is meant by it (in vivo cloning: no; in vitro: yes) was adopted by the Human Rights Convention for Biomedicine. Now the interpreters of the Human Rights Convention on Biomedicine claim that this is not that serious since it is stated in the Article 18.2 mentioned above that one may not produce embryos for research purposes, and this excludes the possibility of using germ cells to create embryos. Yet I have already noted that the concept of 'research purposes' is not yet clear. Is research for health purposes really excluded by this? In the future the courts will seek to clarify this since at present it is not clear from the text. Thus, one cannot assume that there is a total prohibition of human cloning in the Human Rights Convention on

Biomedicine and its supplementary report. Such a prohibition does not exist at all in the framework of the UNESCO declaration, since in the UNESCO declaration only 'reproductive' cloning is forbidden.

The importance of this 'politics of language' becomes clear when taking the fact into account that the representative of the Vatican to the consultation of the UNESCO declaration supposedly supported the prohibition of 'reproductive' cloning strongly without realizing that this was only a partial prohibition. This whole linguistic game cannot be immediately recognized by non-specialists among theologians since they do not constantly have to deal with these matters. In the last opinion of the German Catholic Bishops on the Ethics of Biotechnology the interdiction of 'reproduction' cloning is expressed, maintaining the misunderstanding that all kinds of cloning are included and that 'reproductive cloning' is only a redundant description of cloning as 'reduplication'. This explanation was given to me when I asked for the reason. But the Church of Scotland (Dec. 8, 1998) has seen the problem clearly enough: 'Can we ethically create embryos by cloning – knowing that on ethical grounds we could never allow them to develop to full term?'

This distinction also plays an important role in the cloning prohibition which is demanded by the World Health Organization. There almost identical words forbid the creation of a human being as a copy of another, and this always means 'in vivo'. Thus internationally, there is only one place so far where one may be able to speak of a total prohibition of human cloning. The EU parliament strictly marked its position in a declaration on human rights of January 15, 1998: 'Concerned about the announcement of an American researcher that he intends to clone human beings, with the understanding that human cloning is defined as the creation of human embryos with the same genotype as another distinct living or already deceased human being in his or her total development, beginning with conception, without difference as to the method used, this parliament again confirms that each individual has the right to a genetic identity and that the cloning of human beings must be forbidden without any exception.'

At the level of UNESCO, in the Clinton Commission, and also in the European Union at the level of the European Council in the Human Rights Convention on Biomedicine, and in a declaration of the United Nations (2006), there is obviously overall agreement that in vivo cloning should not be allowed, that is, no cloning that results in the creation of human beings as individual copies. Yet this type of cloning is nonetheless being aimed for. There is, for example, the suggestion of circumventing male infertility by using the man's germ cell together with the woman's ovum and in this way make children possible for a married couple. Richard Seed advocated this as an extension of existing reproductive medicine. Another proposed application is for parents who have lost a child through a traffic accident. They could use one cell of this child in order to clone another one, a 'replacement' which is supposed to bring the parents the joy of repetition. This option is already problematic for purely technical reasons, but also for psychological ones. Such intentions were seriously proposed as ethically irrefutable by ethicists on the Commissions who I know. A third example originates from Philip Kitcher who wrote a philosophical contribution on methods of genetic engineering [2]. A child has kidney problems, for example, and the parents breed a second child in order to have a spare kidney which will not bring with it the problem of possibly being rejected. Germ-cell cloning could be a good method to achieve such similarity between the organs.

The reason for forbidding such options has been the same at every level of political consultation. The ethical principle is the 'non-instrumentalization' of the person. The principle of human dignity according to Kant is: act in such a way that you never treat the other just as a means to an end, but rather always also as an end in him- or herself. Every person affected by our action should thus always also be regarded as an end in him- or herself. This is not a principle forbidding every sort of instrumentalization, but only those which would preserve a person's end in him- or herselfness. This means that a person who has to see him- or herself as the copy of another on request of a third, or any person whose 'design' is affected to conform a third's preferences, would be instrumentalized as such on request of the third, and precisely this would apply to all three examples which I presented. The request of the third person, in this case the wish of the parents, and this is important, does not only relate to the existence. It is not just a matter of a person coming into existence in the sense of wishing for a child of one's own, it is a matter of design; it is a matter of the essence of this person. Or it is an intervention in the essence of the future person in such a way that he or she serves purposes which were externally imposed. At first glance this argumentation seems very convincing, especially in our tradition. But one must not hide the fact that there are strong counterarguments. Let us recognize the Kantian principle: act in such a way that you never regard the other as only a means to an end, but also always as an end in him- or herself. Thus, we are not dealing with an argument against Kant. The claim is made, however, that we are not dealing with total instrumentalization if a second child was created for purposes of transplantation of organ because this child would not be restricted in its moral capacity, which purportedly constitutes human dignity. This second child, after all, would be free, would have the capacity to act, and be capable of moral reasoning. Thus, as long as the capacity to act, that is, the moral capacity, of a future being is not restricted, one can only speak of an instrumentalization that is consistent with human dignity.

Nevertheless, in my eyes it is problematic to identify the anthropological qualities necessary for the capacity for moral reasoning with anthropological qualities necessary in order to respect the other. There are certain anthropological reductions that describe conditions for forming ethical judgments and grounding morality. Among these are the capacity to act, moral capacity, freedom, etc. But these are only reductions, for no human being instantiates them at all times or in its being as such. Thus, being merely reductions, these conditions are not to be directly applied in practical moral reasoning. This would constitute an 'ethicist fallacy'. I am of the opinion that when we speak of human beings as objects of our recognition and regard, we may not transfer the normative reasons for *justifying this* recognition directly to the specific *practical* application, because otherwise those persons who are somehow defective or not yet fully developed in their capacity for freedom, action and morality will easily suffer as a result. Interchanging moral reasoning and the anthropologically determined application of morality to me seem to be wrong.

The prohibition, however, as I tried to show, is not total. It does not include the in vitro prohibition, and with that the question is posed to what extent the argument of non-instrumentalization can be applied in vitro. As far as political consultation is concerned, the supplementary report on the protection of embryos for the Human Rights Convention on Biomedicine of the European Council must still be prepared. The supplementary report on cloning also refers to this missing supplementary report on the protection of embryos. Particularly the discussion in German-speaking countries about the Human Rights Convention on Biomedicine depends on it, and is, after all, not only concerned with this problem.

Theological Arguments on the Ethics of Human Genetics

If one accepts the concept of an autonomous morality in the context of Christian ethics, as many German-speaking moral theologians (e.g. J. Fuchs, A. Auer, F. Böckle) have done for 40 years, then this does not yet mean a decision in favor of a particular philosophical way of thinking in ethics, but only the insight that the basis for judgments is achieved by philosophical means. Assigned to theology is the context of the discovery of ethical values, and obligations contribute significantly to ethical motivation, that is, to the motivation for ethics. Finally, theologically handling ethics – one just has to look at the doctrine of grace and justification – has a very specific function. Theology is relevant to the discovery of ethical problems, to moral sensitizing, to moral motivation and to the relativizing of morally judging persons. In contrast to Kant the epitome of being human is not moral capacity but rather the need for salvation, the 'feeling of absolute dependency', to complement Kant's anthropology with Schleiermacher's. Happiness, freedom, and God are not philosophical postulates, but rather specific religious experiences.

Given this, it is not to be expected that theology will replace, circumvent, or change the ethical arguments in the biomedical and biotechnological debate in general or in the debate about human genetics in particular. This is to be illustrated by an example. In the debate about the protection of embryos, documents of the Catholic Church emphasize that, from the beginning, humans are the image of God, and thus carriers of dignity before God, that is, of personhood (personhood and human dignity are interchangeable in church documents). This doctrine however does not spare us from the need for philosophical reflection in which we must first explain which living entity is meant when speaking of God's image. We may not attempt to replace this reflection either with false biblicism or with false positivism of doctrine – the first is the Protestant, the second the Catholic temptation. This could cause abstruse speculations to come into being. In line with biblicism, for example, the words of the Psalmist, 'You pulled me from my mother's body, since I was in my mother's lap you have been my God' (22;20,11) would mean that being the image of God should be understood as beginning at birth. Or the word of Psalm 51:7, 'my mother conceived me in sin', could be understood to mean that personhood begins at conception. The bible cannot be applied to answer questions which did not exist in biblical times. It is more likely that the way of discourse can serve as a model for solving controversial problems.

A position that regards the fertilized egg as a person for reasons of caution (so-called tutorism) could be reinforced by philosophical arguments about the continuity of the human from the very beginning. But every philosophical argument remains alone, vulnerable and hypothetical, it is never definitive. Only a combination of arguments is strong enough to support the subsumption of the embryo under the principle of human dignity as an absolute value.

If ethical rightness does not belong to the theological deposit (except possibly in that tautological form which prohibits murder, where murder is by definition unethical), then the theological argument can 'only' address ethical insights in contextually anthropological conditioning. It is ethically relevant without making philosophical methods irrelevant. With that, the significance of theology is in no way played down or diminished. The significance of such contextual arguments can be that they change the whole scenery of ethics and, with that, they can significantly influence the relative importance of ethical arguments. In the following I want to illustrate this.

Asked whether there is a relevant theological argument against the cloning of human beings I have indicated, without neglecting the previous explanations, that the diversity of humanity is a commandment of creation. This is Ellen van Wolde's conclusion in the examination of Genesis 11 [3]. The many languages imposed by God are not an answer to human hybris, since towers that reached up to heaven were in no way interpreted as means to assault heaven. (This has been better known among orientalists than among preachers!) It was the urban centralization of humanity, their inclination to uniformity in culture (and religion?) that displeased God. Therefore He reminded them of the commandment of creation to fill the whole earth, and He spread out humanity by the confusion of tongues. Something similar was repeated at Pentecost: the gospel, communicable despite the many languages, was spread throughout the world. Of course, diversity, not repetition of the same, can only be applied as a theological motif if we have good *ethical* arguments in its favor, and we have these as I tried to show above. Just as the argument about the person being the image of God and the argument about dignity are only inspirational in ethics, but are not substitutes, is the thought about diversity also a theological stimulants but not a substitute for arguments.

The stimulating effects of such theological impulses transcend the problem itself. This can be shown by a further theologoumenon of ethical relevance: finitude, that is to say, contingency. Obviously, contingency is also a philosophical concept. Of course, this concept is rather removed from theology when it is combined with 'coincidence', such as in Richard Rorty's postmodern interpretation. In classical philosophy, on the other hand, contingency rather has the meaning: the dependence of being, being time-dependent, finite in the sense of ending at death, capable of mistakes, the imperfection of all that is human. This is exactly what is meant by the theological theorem of the creaturality of humanity – of not being like God. Seen theologically as well as philosophically – both ways are united at Pascal, for example – trying to transcend the own finiteness humanity would gamble away the quality which make up its being human. Theology expresses this narratively in the story of the Fall. Philosophy draws attention to the fact that it is precisely in the corporality of being human where both fulfillment and want deprivation are located (e.g. according to Merleau-Ponty), that finiteness is the form of the possibility of happiness.

Finiteness could be a concept of convergence between theology and philosophy. It is relevant in ethics because it interprets the following as missing the point of what it means to be human; the expectation of humanity's perfectibility, of the principal feasibility of achieving anything, and of being able to still correct all problems that are created a new by problem-solving. This interpretation considerably changes the context of the biomedical and biotechnical debate. It is, for example, a stimulant for the ethical rule of problem solving: one should not solve problems in such a way that the problems which arise through the solution are greater than the problems which are solved. I do not believe that the paradigm of progress in modern biotechnology, including human genetics, keeps to this contextual insight. In biopolitics I have observed that in this regard continental Europeans, even if they are agnostics, can be convinced of this insight more easily than US Americans, even if they consider themselves to be Christians. How is this possible?

In conclusion I want to introduce a fourth theological argument – after being made in the image of God, creation's commandment of diversity, and of finiteness – which I often painfully experience as missing context of reflection: the vulnerability of humanity. Analogously to the theological 'option for the poor' in our context (biotechnology, human genetics) it is possible to formulate an option for the priority of vulnerable persons. Here one easily runs into a lack of understanding in the secular context. Certainly, the average pragmatist who does ethics in political advisory groups will acknowledge that one must respect disabled persons and incompetent persons, even (more or less) fetuses and embryos. But he or she will face a theological theorem such as 'the prophecy of disabled persons' with little understanding. And yet, still it is easy to explain to him what it means: being explicitly disabled reminds us that we are all disabled in some way. It reminds us of our finiteness.

In the practice of finding ethical judgments in an accelerated world of suspended doubt, one often meets the readiness to represent persons who are not able to realize their interests and opinions by themselves, that is, those who cannot articulate themselves, by extrapolating their presumed intention via analogous deduction. The fiction of 'informed consent' is stretched to include them. Curiously, exactly by this they tend to be instrumentalized. When, for example, the Human Rights Convention of the European Council on Biomedicine tried to formulate the conditions under which research for the benefit of others can be carried out on those incapable of consent, it forgot to secure in the document the ability to withdraw proxy consent, as was discovered by my Protestant colleague from Vienna, Ulrich Koertner. A whole mindset can be recognized in this forgetfulness!

There is an aura of vulnerability around persons who cannot represent themselves, the luster of the butterfly that may not be touched. Forgive the poetic expressiveness, but I want to illustrate that we are dealing here with the need for greater sensitivity. Sensitivity can be easily declaimed. We then carry the burden of proof that it is not present. Our time lives with the ideal of the person as young and strong, an ideal for which Paul already admonished the Corinthians. It is not just a matter of developing a culture of perception and feeling (although also of this). It is a matter of the role of taboos which have grown out of historical experiences, for example the Nurnberg code which simply forbids experimentation on those not capable of self-determination. Taboos can only be ethically justified if they are understood as cautionary fences. They do not belong in an ethical justification but in its context which modifies the importance of the justification. Or stated differently: Situations of vulnerability demand caution. I keep asking myself why the Bioethics Convention of the European Council did not allow advocates for the affected groups themselves to formulate the norms regulating non-therapeutic research on those incapable of granting consent. Here, a sense of superiority of the experts is expressed which to me seems embarrassing.

The reticence of the Catholic Church in this regard puzzles me. The church either has not reacted, or reacted with cautious agreement, to the international advisory texts, the declarations (UNESCO) and conventions (European Council). On the other hand it disciplines theologians and the faithful for reasons that are difficult to defend. Even the suspicion of ambiguity in pregnancy crisis counseling (obligatory counseling in the 3 months during which abortion is not penalized), which could (but de facto does not) fall onto church-related counseling services, has lead to vehement interference with the German bishops by the Pope and the Prefect of the Congregation for the Doctrine of the Faith. Not even a stain of merciful practices is to fall onto the clean, white teaching robe of the church's interdiction of an abortion which is penalized by excommunication. On the other hand, church authorities hardly, or only very indirectly and abstractly, criticize possible mis-developments of biotechnology, that is to say, human genetics. Should the Catholic Church, in opposition to the intent of the Pastoral Constitution of Vatican II, move in the direction of developing a split between the message for the world and the message for the church? The inclination only to pay close attention to ethical problems when they are ecclesiastical problems emphasizes the march into the Catholic ghetto more than it would happen by a well-argued critique of the 'modern' system of interdependence of research, technology, and economy. However old and wise the church authorities may consider themselves to be, the discussion on 'Bioethics' cannot be resolved by declamations like 'Donum Vitae', a Declaration of the Roman Congregation of Faith (1987), for which a repetition is foreseen for December 2008. On the other hand, the interest in protection of a personhood, which not only depends on actual human conscience but on human existence in itself, may also be the attention of the public not only the church.

The Anthropological Question: The Ethical 'Status' of the Early Embryo

The notion 'status', used nearly everywhere in this context, is nevertheless not self-understanding. It may be that the biopower in this case anticipates the intention in the key word 'status'. A status is always given. Is the 'status' in its context given by 'nature'? 'Nature' however does not behave or reason according to ethical standards, nature is not respecting or actively protecting embryos. Is the 'status' given by culture? This is also problematic if one remembers that in the Roman Empire the father had the right to acknowledge a child for its right to survive. Is the 'status' given by philosophical reflections? But can any reflection 'give' someone human dignity and human rights, especially if he or she is already essentially a human being? The notion 'status' could be regarded as merely a remainder received from the Christian faith, in which indeed the 'status' is 'given' by creation. But the gift of creation could not be used in a discussion about the anthropological question who belongs to the human beings. This is easily empirically verified. Is there any alternative?

What is a fertilized egg cell, an early embryo in an anthropological (philosophical) perspective? The European Convention on Biomedicine offers three possible answers which are mentioned in the supplementary protocol on human cloning: a person, a human being, and a conglomeration of human cells. The Embryo Protection Protocol, which was foreseen among the four protocols applying the Convention, was never decided by the General Assembly of the Committee, only a Report on Embryo Protection by the Working Group GT3 was declassified in 2003.

I suggest that the last of these three possibilities is excluded in this report, because there is indeed a significant difference between human germ cells, such as egg and semen cells, and an early embryo. The embryo has a gender! It is a fact that the embryo has the ability – and not merely in the sense of an abstract potential, but rather in the sense of a real capacity – to become a human being if its development follows its inherent intention, that is to say, when nothing is undertaken or omitted to the detriment of this development. One might object that 'nature' does not implant all fertilized egg cells. But this objection is misplaced in this context because 'nature' cannot be treated as a morally responsible subject at all. Whoever is not prepared to accept 'nature' to establish an ethically restrictive argument should not attempt to use it for a normative argument for indifference either.

If, however, the embryo must be conceived as a human being, then we need at the very least to speak of it having a morally relevant status. Can one maintain on the basis of this status that every embryo is the bearer of individual rights which exclude its destruction or even its being put at risk? Those who do not want to protect embryos individually, but would rather protect them only as a particular kind of 'biological material' which is to be treated with respect by means of allowing their use only to a particularly small and limited extent, in my opinion already violate the above-mentioned morally relevant status of a human being. My point of view assumes, of course, that belonging to the species 'human being' already entails a particular right to protection which transcends that for animals.

But is the issue not larger than this? The Catholic position proceeds from the assumption that an embryo is to be treated 'like a person'. This formulation is carefully worded to the extent that it does not simply maintain that embryos are identical with persons. It is argued, of course, that because of the unity of human development and because of its continuity, which cannot be broken down into really different phases by even the most major stages of development – and indeed because of the unpredictable consequences it would have for other areas if one began to distinguish between human beings on the basis of this kind of personal quality – one cannot make a distinction between 'human beings' and 'persons' and assign them to two different levels. The main argument here is thus the continuity of being human and its inseparability. The consequence is that a moral status emerges which grants the embryo a fully valid protection of life and does not allow research on embryos which treats them as raw material. If one holds on to this status, then life as a fundamental right cannot be weighed up against other highly ranked goods either.

This position, which is itself philosophically argued – in the absence of statements of revelation in this area which can be proven to transcend a particular time and place, a theological position can only be philosophically argued – is also attacked philosophically. Without discussing here the relevant arguments in detail, let me say that a notion of personhood which is tied to cognitive qualities certainly does not correspond to the Christian tradition, because this tradition argues either in terms of substantial or structural ontology (and I mean by this the capacity of the structure), but in no sense in terms of epistemology.

This position also excludes the possibility of what has become common in the indifferent pragmaticism of the participating sciences, namely, that a distinction is made between embryos with an assigned bonus for implantation, and embryos with an assigned inadequacy for research. With this kind of decisionism, one spares oneself any need for further ethical reflection.

There is a tendency among the supporters of research using embryos as raw material to make absolutes out of the medical options and the only potential and in no sense certain therapeutic benefits. Research, however, uses up material now and irrecoverably. The interests of future sufferers from an illness are important, but they must not be made into absolutes in a society which needs to be committed to all the relevant values. The social solidarity which is expressed in the resources placed at the disposal of research needs also to set priorities within questions of solidarity. In question is also the solidarity between women and with women. For this question we take as an example the so-called 'freedom of procreation'.

The So-Called 'Freedom of Procreation' and the Instrumentalization of Women

There are two feminist approaches to IVF, preimplantation genetic diagnosis (PGD) and connected technologies. One of them is to be seen in the context of political liberalism. In a rather individualistic approach, women who realize and defend their rights in an androcentric society, are striving for a good or better life, i.e., in their view, for more freedom of their choices in all fields of societal and private life. In the field of human reproduction the pressure on the self-determination of a woman, i.e. of 'nature', the pressure of the predominance of men in setting up regulations, the pressure of antiquated authorities, etcetera, seem to be very strong. Women's liberation in this line of arguments means to reclaim freedom of procreation as negative freedom: absence of any kind of manipulation or compulsion in procreative decisions. Insofar as the future embryos and fetuses are included in this freedom, the imagination is that, if it must be decided over their lives, the future mother will be the best decision maker in favor of the future child's life and of the quality of this life.

In solidarity with this approach many people involved in biomedical ethics, especially medical doctors, often reclaim the self-determination of women as the moral key for a morally right decision. The principle of informed consent and the purpose of adequate, nondirective counseling seem to be in a good correspondence with this approach.

Taking into account the distinction generally made with respect to human rights, the question is whether the 'freedom of procreation' as a woman's right is claimed as a more defensive – negative – right or as an offensive – positive – right. In the first case, a woman must be protected against interventions which interfere in the process of reproduction without her consent. But the question remains if every single part of reproduction is subjected to the arbitrary option of every woman. This question can be answered if a positive right is admitted. The positive right to 'freedom of reproduction' includes another kind of corresponding duties or solidarities than the defensive right. If the free choice includes all methods of *assisted* reproduction which are available and these are the issues of the positive right, then society is obligated to promote all questionable methods and to make them available to requesting women.

The starting point of women's rights to free procreation is understandable because in the biographies of women the problem of giving birth to a child early in her life or having a child with severe health problems is a great risk for self-realization and can even turn into a risk of poverty. So, the expectation that modern reproductive methods can solve all the different kinds of problems accompanying the woman's striving for a good life by a perfect biomedical organization may become widespread, but nevertheless illusory.

In contrast, the negative right to 'freedom of reproduction' does not include the obligations to make all methods of assisted reproduction available. Furthermore, negative rights arguments do not even implicate a preference for IVF or intracytoplasmatic sperm injection methods, since they are only addressing the issue of noninterference.

On the other side: the feminist concept of self-determination cannot be reduced to free choice with respect to the supplies of biomedical progress in the reproductive area. It is illusory to believe or claim that the problems of living a good life and quality of life in the sense of social qualities can be resolved by biomedicine. Often, in this regard the success rate of IVF is quoted as a remaining negative factor. But it is not only the success rate which must be considered with respect to the good life women are striving for. Beyond that, IVF treatment includes a lot of risks and complications for the women as well as for the future child. Psychological studies have shown that there is often a lack of information and transparency. The risks of late abortion and the necessity to undertake IVF in order to realize PGD (with the necessity to control the preimplantation diagnosis by PND) are one of the reasons why more and more the other feminist approach comes into the discussion.

This other feminist approach which has been part of the feminist reflection right from the beginning of modern assisted reproduction, takes account of the way in which reproductive medicine can be used for the discrimination and instrumentalization of women. This is, above all, seen with respect to the promotion of research in embryology and reproduction, with respect to the relief or suspension of physicians' responsibilities and with respect to the omission of social reforms which are necessary to make a good life possible which includes the option for procreation in every age and in every economical situation. Thus, this approach points more to the women's risks in assisted reproductive methods, and the question of unsuccessful treatment is becoming more important if balanced against the possible risks and the overall constraints in the procedures. Critical remarks are made on discriminating or even abusing women and their bodies for an androcentric conception of research and for economical interests, for instance in the question of delivering eggs for different purposes and in the question of making women's bodies disposable. Instrumentalization and the disposing of women's bodies and lives through assisted reproductive procedures thus become the two main critical points of this feminist approach. In contrast to that, these feminists strive to improve the conditions of women by reducing the social pressure on their options for a good life with more social solidarity in order to be able to give birth to a child at an earlier age, to balance reproduction and professional career, and, more generally, to change the gender-biased ascription of parenthood to the disadvantage of women.

In the embryonic stem cell discussion another point may be underlined: To gain spare embryos, women have to undergo IVF treatments. This means that with embryonic stem cell research, the interest of third parties will gain influence in reproductive medicine. This may be risky because of the vulnerability of women undergoing

IVF. In the case of so-called therapeutic cloning, to produce one embryonic stem cell line by using cloned embryos, hundreds of donated egg cells are needed. This critical approach is mentioning that the need for so many egg cells will lead to a dynamics which will endanger the respect of women's rights. Can women be degraded to deliverers of raw materials for research purposes?

Over the last years, the Western culture-dominated feminist approaches have been complemented by a more global perspective, taking into account that in most non-Western countries women are discriminated against men in the access to healthcare. This perspective sheds an important light on the attempts to spread assisted reproductive techniques as well as genetic diagnosis (PND or PGD) to those countries which do not guarantee women access to methods of birth control or, even worse, to any kinds of basic healthcare. Furthermore, the issue of 'social' sex selection has become a major topic of feminist concern.

References

1 Lenoir N: in Miltner F: Hoffnung und Horror. Focus 1998, No. 8.

2 Kitcher P: The Lives to Come: The Genetic Revolution and Human Possibilities. New York, Touchstone, 1996.

3 van Wolde E: Words Become Worlds. Semantic Studies of Genesis 1–11. Leiden, Brill, 1994.

Further Reading

Bentele K: Ethische Aspekte der regenerativen Medizin am Beispiel von Morbus Parkinson. Mensch-Ethik-Wissenschaft, Bd. 4. Münster, Lit, 2007.

Cahill LS (ed): Genetics, Theology and Ethics. An Interdisciplinary Conversation. New York, Crossroads, 2005.

Duewell M, Rehmann-Sutter C, Mieth D (eds): The Contingent Nature of Life. Bioethics and the Limits of Human Existence. International Library of Ethics, Law, and the New Medicine, vol 39. Berlin, Springer, 2008.

Englert Y: Gamete Donation: Current Ethics in the European Union. Oxford, Oxford University Press, 1998.

Ford NM, Herbert M: Stem Cells: Science, Medicine, Law and Ethics. Strathfield, St Pauls, 2003.

Haker H, Hearn R, Steigleder K (eds): Ethics of Human Genome Analysis. Tübingen, Attempto, 1993.

Haker H, Beyleveld D (eds): The Ethics of Genetics in Human Reproduction. The Status of the Embryo 59–100, Legal regulations 215–294. Burlington, Ashgate, 2000.

Haker H: Ethik der genetischen Frühdiagnostik. Paderborn, Mentis, 2002.

Hilpert K, Mieth D (eds): Kriterien biomedizinischer Ethik. Theologische Beiträge zum gesellschaftlichen Diskurs. Freiburg, Herder, 2006.

Hoedemaekers RHMV: Normative Determinants of Genetic Screening and Testing. Wageningen, Ponsen en Looijen, 1998.

Honnefelder L, Lanzerath D (eds): Cloning in Biomedical Research and Reproduction. Scientific Aspects – Ethical, Legal and Social Limits. Bonn, University Press, 2003.

Junker-Kenny M (ed): Designing Life? Genetics, Procreation and Ethics. Aldershot, Ashgate, 1999.

Kaiser G, Schmidt S (eds): Stammzellen und therapeutisches Klonen – Biomedizin ohne Grenzen? Düsseldorf, Das Magazin, Wissenschaftszentrum Nordrhein-Westfalen, 2001, No. 2.

Kersten J: Das Klonen von Menschen. Eine verfassungs-, europa- und voelkerrechtliche Kritik. Jus Publicum 115. Tübingen, Mohr Siebeck, 2004.

Koertner UH, Kopetzki C (eds): Stammzellforschung. Ethische und rechtliche Aspekte. Schriftenreihe Ethik und Recht in der Medizin, Band 2. Berlin, Springer, 2008.

Mieth D, Pohier J (eds): Ethics in Natural Sciences. Concilium No. 203. Edinburgh, T&T Clark, 1989.

Mieth D: Was wollen wir können? Ethik im Zeitalter der Biotechnik. Freiburg, Herder, 2002.

Mieth D: Stem Cells: The Ethical Problems of Using Embryos for Research. J Contemp Health Law Policy 2006;22:439–447.

President's Council of Bioethics: Human Cloning and Human Dignity. An Ethical Inquiry. Washington, Public Affairs, 2002.

Rehmann-Sutter C, Duewel M, Mieth D (eds): Bioethics in Cultural Contexts. Reflections on Methods and Finitude. International Library of Ethics, Law, and the New Medicine, vol 28. Berlin, Springer, 2006.

Voeneky S, Wolfrum R (eds): Human Dignity and Human Cloning. Leiden, Brill, 2004.

Dietmar Mieth
Blumenstrasse 3
DE–71149 Neustetten (Germany)
E-Mail dietmar.mieth@uni-erfurt.de

Pfleiderer G, Brahier G, Lindpaintner K (eds): GenEthics and Religion. Basel, Karger, 2010, pp 101–107

First Sheep, Then Human Beings? Theological and Ethical Reflections on the Use of Gene Technology

Eberhard Schockenhoff

Theologische Fakultät, Universität Freiburg, Freiburg, Germany

The development of modern gene technology makes it possible for human beings to go beyond limits which were established in the previous course of their history. As long as the scope of their scientific and technical activity on nature was limited by nature itself, the question of the legitimation of such interventions did not arise with the acuteness which it now has. We must ask about the human significance of limits which were formerly imposed by nature, and in religion, ethics and law look for sustainable foundations on which to establish new limits. What technology can do does not automatically satisfy the criterion of ethical responsibility and what is humanly desirable.

Limits of Nature – Limits of Ethics?

The faith in biblical creation reminds human beings both of their commission to rule and of the special position which this gives them among all creatures, and also of their inalienable responsibility for the wellbeing of the whole of creation, which accrues to them precisely as a result of their exalted role as God's representatives and governor. Their charge to shape creation and their place among the creatures thus belong inseparably together: on the basis of the similarity of all creation to its divine origin, nature too gains a part in the reverence which human beings should feel towards their transcendent Creator. But in so far as nature not only reflects the beauty and glory of the triune God and the divine fullness of life, at the same time this is distorted as a result of the breach caused by sin, which in fact cannot simply be identified with the divine will for creation.

If at the same time according to biblical understanding, the origin of evil is not to be sought in a negative principle of the world or in the dark aspects of the deity, but in

Translated from German by John Bowden.

a contingent state of freedom referring back to the first beginnings of human history, the consequences of evil materialize in the structures of the fallen creation. They are tangible in the destructive forces and the disruptive tendencies which threaten both the life of human beings and animals and inanimate nature in the form of diseases, accidents and catastrophes. Because creation is always also the place where evil breaks in and never exists only in the undisturbed reflection of its divine origin, according to the biblical understanding, in principle human interventions are allowed to sustain nature and make it useful; theologically they must be understood as collaborations with the divine action in creation. Since the limits of such interventions in an evolutionary dynamic picture of the world cannot be read off existing ordinances of nature, they must be rediscovered and recognized through being opened up by human reason. That there must be such limits follows from the finitude of human beings and the limitations of their constitution. However, precisely where these limits run cannot be discovered in advance of the state of possibilities of scientific and technological intervention at any one time. Rather, critical reflection is needed on the conditions and consequences of human action which investigates what technology is capable of at any time on the basis of what is reasonable for human beings. Thus scientific self-control and ethical reflection are called for to define afresh the limits within which science and research, medicine and gene technology, serve human wellbeing.

The criterion of human dignity, respect for the life of others, and the prohibition against killing, give us important criteria for forming ethical judgments which make it possible to define the first limits in situations of conflict in bioethics. Common to all these criteria is that they formulate only a negative restriction, and not as yet any comprehensive aim. They describe the limiting conditions on human action in the borderland between life and death.

Below this level, at which the claim of the person is expressed in negative obligation relating to what is not to be done, we need further ethical criteria of evaluation so that we can assess particular research projects in gene technology and the procedures used in them. Here above all we must consider the justification for the aims and the responsibility for the consequences; this must precede or accompany individual research projects. In connection with the present possibilities of using gene technology on human beings, here we shall investigate the opportunities for forming an ethical judgment by means of two examples. In one case (genetic diagnosis) the ethical reflection tends to be retrospective and to go along with scientific progress, whereas in the other case (cloning) the future effect of its possibilities is anticipated.

The Right Not to Know

In principle, the judgment to be made on the methods of prenatal diagnosis is no different from that on the procedures of medical investigation known so far. The ethical problems associated with it are not limited to the danger of a miscarriage or damage to the embryo in connection with taking tissue, although this risk attaching to intervention must also be included in the assessment and in the counseling of the parents, since in the invasive procedures practiced at present the risk of damage is approximately equal to that in the diagnosis of a disruption of chromosomes (in amniocentesis it is between 0.5 and 1%, in chorion biopsy between 2 and 4%). The real problem in antenatal diagnostic procedures lies in the practical consequences of their application. It is the medical aim of prenatal diagnosis to recognize at an early stage developmental damage with a probable or (in conditions with a monogenetic cause) certain prognosis, in order to dispel any existing anxieties of parents and make possible the best medical care both for the mother and the unborn child. In more than 97% of all cases prenatal diagnosis leads to a negative finding; thus in numerous pregnancies in which there is a risk, it can make it easier to accept the child and thus prevent the pregnancies being terminated because of the vague fear of possible chromosome damage to it.

The ethical dilemma to which the use of prenatal diagnostic procedures can lead becomes evident in the relatively rare cases in which the prenatal test predicts a genetic anomaly in the child. As long as there are only a few established medical therapies and the operative techniques are still at an early experimental stage, the diagnosis of a risk of genetic damage exposes the parents to the pressure of mental expectations where the thought of abortion is widespread, a pressure which only a few find the courage to resist. The ethically legitimate desire for a healthy child and the medical aim of avoiding hereditary illnesses then lead to the decision to have an abortion being taken as an obvious consequence on medical grounds, and being derived from a catalogue of supposed forms of damage to the embryo.

According to the view which nowadays is largely taken for granted, the care of a handicapped child and life with it call for more personal readiness for sacrifice than law and morality can ask of an individual. This presupposition, backed by an uninvestigated consensus in society, is evidence of a deep inner contradiction which underlies the possibilities of prenatal diagnosis and genetic counseling of families. Whereas from a medical perspective the aim of this counseling is to be able to recognize and treat ailments in the developing child as early as possible, the secret expectation in society is one of the increasing suppression of the birth of handicapped children in the future with the aid of the diagnostic techniques of molecular biology. If prenatal diagnosis can be extended to an increasing number of ailments or even to genetic characteristics with only a small element of disease, this will also encourage the further development of eugenic tendencies in the population. Even now, in surveys on the question more than 40% of those surveyed say that they regard a genetic disposition to excess weight as sufficient indication for a precautionary termination of pregnancy.

The routine use of general screening procedures can lead to a change in the assessment of handicap and illness in the consciousness of the population and a tendency to regard any deviation from 'normality' as an intolerable limitation. To the degree that medicine is expected to fulfill the hope of the birth of a healthy child, and the diagnostic procedures of molecular biology are understood as sure 'guarantees' of this, the social resistance to and discrimination against the handicapped will increase. There is unmistakably less readiness to accept handicapped children, and they are seen as a lifelong task which can bring not only limitations and burdens but also enrichment of one's own life.

Since the limits of one's own strength cannot be assessed in advance with any certainty, and the actual degree to which burdens can be tolerated depends very much on the personal attitude which parents develop even before the birth of their child. Early knowledge of a risk of illness weakens the capacity to accept a handicapped child in the same way as a healthy child which, at first sight, seems more to satisfy the wishes of its parents. Moreover, the inability to imagine life with a handicapped child in the tangible burdens on everyday life as a serious alternative is reinforced by anxieties about public reactions. There is a fear that later one will have to justify oneself before society for the mere existence of a handicapped child, the acceptance of which one could have spared oneself and the community. Whereas formerly, after the chance birth of a handicapped child, parents could still count on the sympathy of neighbors, friends and relatives, in future they will increasingly come up against misunderstanding and rejection as a result of their decision voluntarily to accept a child which could also have remained unborn.

Even if such dangers of misuse, which are extremely difficult to control, do not rule out prenatal diagnosis, they must not be banished from social awareness. From an ethical perspective, there needs to be strict observance of three limits in particular which threaten to be forgotten in the uncontrolled use of investigative techniques based on molecular biology. In principle, prenatal diagnosis may be used only within the framework of medical questions: the discovery of general characteristics like gender, phenotypic characteristics or even psychological dispositions, should this ever be possible with a sufficiently certain prognosis, are ethically indefensible. Even if genetic tests had to be offered to all parents, no one should be forced actually to take them. Even the use of such prenatal diagnosis when indicated medically needs the assent of the parents, which can be given only after a comprehensive explanation of the risks of miscarriage or damage to the child, and the moral conflicts which could arise were the finding to be positive. Even a refusal to employ prenatal diagnosis is ethically defensible. Indeed it calls for special respect and esteem, where there are no clear signs of a substantial genetic handicap in the family. The special ethical value of such an attitude lies, first, in a protest against an illusory assignation of tasks to modern medicine, and secondly, in the readiness which it indicates to recognize human life without reservations, without subjecting it to a discriminatory assessment.

The Right to Natural Birth and Genetic Chance

If human interventions in nature are in principle legitimate, then the cloning of animals cannot be objectionable simply because it goes beyond the limits which have been applied so far. Nor can the novelty of the procedure be the basis of a categorical prohibition. However, because there is a need to preserve the multiplicity of species and respect the intrinsic value of animals, the fact that cloning is permissible in principle must not be understood as an unlimited authorization which legitimates the use of cloned animals for any purpose that is desirable from a human perspective. The techniques of splitting embryos (which have recently been tested) or the transplantation of nuclei into previously denucleated embryonic cells from which their own genetic material has been removed, can only be justified in any balance between the benefits and the animal's own interests, where from a human perspective it can be said that the goals of research are important and that the interests in terms of health are particularly urgent.

Among the conceivable fields in which cloning can be used in the foreseeable future and in which human beings can meaningfully make use of cloned animals, it will be necessary to make different ethical judgments depending on the particular aim. There is doubtless greater justification for so-called 'gene farming', which is meant to guarantee a consistent quality of drugs and important pharmaceutical substances and for the transference of animal organs to human beings, than there is for an unlimited extension of the quantity of high-quality meat production, once that has been achieved, which is solely in the human interest. Since the animals of which human beings make use also have a creaturely value of their own over and above the functions which they fulfill, in the future, too, they should be preserved in all their genetic multiplicity and not be reduced to the function of cloned high-yielding species, bioreactors which can be reproduced at will and be substitutes for pharmaceutical machines. By contrast, the cloning of human beings should that ever prove successful, is incompatible with the notion of the personal dignity and individual purpose of each human being. The applications which are hoped from a transference of cloning techniques to human beings – at present mention is made, for example, of improved methods of treating sterility, the avoidance of genetic hereditary diseases in families at risk, the possibility of multiplying the best transplant material for organ donors at will – are in part defined as medical research interests of great importance, which it can be ethically legitimate to pursue. However, in other cases the aims mentioned cannot be justified, nor are the consequences expected responsible. This is true in particular of the fantastic speculations about the infinite revival of dead human geniuses or the replication of individuals who are regarded as outstanding for the multiplication of twins at different times, or an increase in the production of embryos within the framework of embryonic research.

Independent of the question whether cloning can be legitimate at the level of goals, the cloning of a human being must already be regarded as intrinsically reprehensible. An individual who came into being in this way would not be wanted for its own sake; it would be produced only as a means to an alien end. To force someone else's genetic identity on a person and to want to multiply someone because of the properties, capabilities or characteristics to which his or her genomes disposed him or her amounts to a clearly recognizable instrumentalization which is incompatible with human dignity and the notion that human beings are an end in themselves. A human being who is to be brought to life only as a genetically appropriate organ donor, as a replacement for a spouse who died prematurely, or as a reduplicated twin of its parents' eagerly awaited only child, could not exist as an end in himself or herself: such a person would have been affirmed only for a purpose external to his or her interests and would be accepted by those around only because his or her genetic characteristics served a particular purpose. Since the mere decision to produce such a person was made only for an alien purpose or to achieve the aim of a third party, from the first moment of his or her existence he or she would be deprived of the possibility of leading a life which had not been lived before and of being respected as an identity which was not under anyone's control and, in this sense, was 'fortuitous'. In analogy to the right to ignorance which human beings have so that they cannot be forced to know the risks to life inherent in their genetic and biological make-up, a right to individual fortuitousness must be postulated which protects the human being from the planned intervention of his or her procreators.

Here the crucial issue is not just that a human being has control over the genome of another, or that a number of human individuals share one and the same genetic makeup. The personal individuality and inalienable uniqueness of a human being do not end in his or her genetic programming: rather, the development of personal identity takes place in the process of an open interchange between natural disposition and environment which is only partially determined by the basic genetic program. Therefore the human genome cannot be seen as the exclusive vehicle of the personal identity of a human being. However, the basic genetic makeup does provide the natural framework for the development of the future personal identity of the human being. Along with other determinants of our physical existence, the individual genome prescribes the biological scope of moral freedom, which on the basis of the psychosomatic unity of the person is the antechamber to the idea of human dignity and, because of the elementary possibility of the responsible planning of life, must be included among the areas which need protection.

The right to the chance origin of one's own existence from a process of procreation in which two biological parents are involved, which is called for in the documents of the church's magisterium, therefore does not follow from a particular dignity of the process of procreation which already in itself has moral dignity and inviolability. The right of the human person to be produced by his or her parents in a natural way is grounded, rather, in the fact that only the chance combination of two haploid nuclear cells to form a new individual human genome provides protection against becoming the object of genetic manipulation and heteronomous determination by an outside party. The objection that the biological parents also determine the genetic identity of their children misses the decisive point, as in the process of procreation these are acting as natural beings who have no deliberate influence on the composition of the genetic identity of their children. Nor is the reference to the occurrence of monozygotic twin births in nature a tenable objection to the argument which has been put forward. Unlike human clones, twins are not identical in the sense that all their hereditary information was determined by a single alien individual. Certainly they have a common genetic heritage, but this does not derive from a single genetic copying instrument but from two biological parents who because of the chance nature of the process of procreation can never again produce a genome of precisely the same nature. Moreover the process of cloning can be repeated endlessly, so that in this way any number of identical examples of the original genome could be produced, and that does not happen in nature.

Above all, however, monozygotic twins differ from clone and donor by virtue of the contemporaneousness of their existence. They come into existence together, and both have their future ahead of them independent of each other, whereas clones would always be dictated to by expectations and predictions which have already been played out in the displaced lifetime of their predecessors. Here we must remember with Hans Jonas that human beings can be really free only if they do not know their future and their own fate, whereas clones, which are displaced in time and succeed one another to such a degree that in a hypothetical thought-experiment we could put ourselves in the situation of such beings, would be condemned to a monstrous existence without any natural quality and spontaneity.

Eberhard Schockenhoff
Theologische Fakultät, Universität Freiburg
DE–79085 Freiburg (Germany)
Tel. +49 (0)761 203 2089, Fax +49 (0)761 203 2096
E-Mail Eberhard.Schockenhoff@theol.uni-freiburg.de

Pfleiderer G, Brahier G, Lindpaintner K (eds): GenEthics and Religion. Basel, Karger, 2010, pp 108–117

Is the Human Genome Sacred?

Ted Peters

Pacific Lutheran Theological Seminary and Center for Theology and Natural Sciences, Graduate Theological Union, Berkeley, Calif., USA

Why might we pose the question: *is the human genome sacred*? Because it carries a potent cultural valence. Like a sacred object, our genetic code has become identified with our essential being. The genome appears like something sacred because it manifests itself as a fundamental reality, as a determiner of our nature, as a special gift from our mysteriously long evolutionary inheritance, as something inviolable, as something to be protected from the profanity of technology. Our genome is not like our hair; we cannot trim it or color it and still retain our identity. If we believe our genome constitutes our essence, then to alter it by technology would be to lose ourselves in an evolutionary oblivion. The very decision to engineer our genome or to mix our genome with that of an animal would be tantamount to Prometheus stealing fire from the sun. It would signal human hubris, a crossing of the line that divides what is heavenly from what is mundane. It would warrant that the gods of Olympus retaliate with punishment. Human hubris expressed in the form of genetic alteration would require punishment in the form of a natural disaster, a disaster of our own making so fearsome that we shrink from contemplating the consequences.

We are working here with a triangle of forces: science, culture, and theology. The laboratory scientist has given us naked data about the human genome. But our culture has draped our genome with existential meaning. Much more than merely a chain of chemicals, our DNA has become identified with the ground of human being in general and of our individual personhood in particular. Having observed this blending of science and culture, the theologian needs to distill the blend and separate the ingredients. Then, the theologian will be in a position to ask whether to respond to the science alone? To the culture alone? Or, to both together?

Armed with a theological anthropology, the theologian can observe here a cultural propensity to over-dramatize the unique role our genome plays in making us who we are. This over-dramatization leads to a vague form of genetic determinism. This determinism slips quickly into essentialism. With essentialism arises a secular variant of sacredness. This sacredness leads gradually to a prohibition against manipulating DNA for engineering or other purposes. This observation is worthy of analysis and comment. The human genome, as vital as it is for human definition, belongs to the created order. It is not divine. Nor is it sacred. Nor should we put up ethical stop signs to prevent intervention, alteration, mixing, or even enhancement.

A Rumor Out of Control

The assumed sacredness of DNA seems to be a rumor out of control. First whispered by scientists two decades ago in support of the Human Genome Project (HGP), it has spread through culture so that we now assume that 'it's all in the genes'. The human genome – the composite of all the genes in human DNA – has been commonly described as the *blueprint* for what is human. At the beginning of HGP in the late 1980s and early 1990s, Kevles and Hood [1] referred to the genome as the 'code of codes'. Walter Gilbert likened the search for the human genetic code to the search for the Holy Grail [1, p 96]. In *Time* magazine James Watson was quoted to have said, 'We used to think our fate was in our stars. Now we know, in large measure, our fate is in our genes' [2]. With such exuberances by renowned scientists, how could our culture avoid building altars to the about-to-be-revealed holy within our biological makeup?

Commentators and critics have suggested that genomic scientists overstated the decisive role played by genetic information. This has led inadvertently to what Nelkin and Lindee [3] have labeled, 'genetic essentialism'. Maddox [4] called it the 'strong genetic principle'. Hubbard and Wald [5] have chosen the term, the 'gene myth'. Of these, I have elected 'gene myth' as the term that best illuminates what is happening at the intersection of science and culture.

The core of the gene myth is made up of three doctrines: (1) puppet determinism, that is the assumption that genes are determinative of who we human beings essentially are – 'it's all in the genes'; (2) Promethean determinism, that is our genetic scientists can reach into our genomes with wrenches and screw drivers to alter our genetic code and give direction to the future of human evolution, and (3) the fear that such Promethean hubris will precipitate a backlash in nature that will lead to some form of destruction. These three doctrines lead to the ethical injunction: our scientists should not play God with the human genome. This set of assumptions constitute a myth, I contend. The gene myth frames scientific research; but, in itself, it is not scientific. The extra-scientific reach of the myth leads our culture to think of the human genome as sacred, as a point of ethical departure for moral guidance [6]. Theologians who are habitual interpreters of myths from biblical times should apply their skills to the interpretation of the Promethean myth alive and well in our contemporary culture's interpretation of the human genome.

Are We Really Playing God?

As just mentioned, the gene myth has given rise to the commandment: *thou shalt not play God?* What is noteworthy, I submit, is that this commandment does not arise from Christian theology, nor from any other contemporary religious tradition I can identify. Rather, it is a secular commandment. It arises in response to biological sacredness. It cautions us against profaning what is sacred, namely, our genetic code.

What does it mean to play God? It means to *alter life and influence human evolution.* What we moderns fear about science and technology is that they will team up so that the scientific understanding of nature can lead to our technological control of nature. Yet, we wonder: should we really try to control nature? Has nature in the modern world replaced what was sacred in the ancient world? If so, will nature rebel against our scientific and technological attempts to arrest it, harness it, engineer it, press it into our service?

As Prometheus was punished by Zeus for stealing fire from the sun, so also we fear that nature will punish us if our scientists enter the holy realm of DNA to steal its secrets and profane what is sacred. Our culture lives with the barely articulated fear that human hubris or pride – the exalted and unrealistic assumption that human technology can remake nature according to our own desires – is the tragic flaw that will lead ineluctably to humanity's self-destruction. The 19th century Frankenstein legend elicits the fear of playing God when it applies the ancient Promethean myth to the scientific age; and Michael Chrichton funnels it into the gene myth with his science fiction novel, Jurassic Park. Leon Kass, former chair of George W. Bush's President's Council on Bioethics, puts words to his incipient yet widespread fear of 'the Frankensteinian hubris to create a human life and increasingly to control its destiny; men playing at being God' [7].

The gene myth does not remain strictly secular. It spills over into religious traditions. Some religious respondents have incorporated the gene myth into their anthropological and ethical schemes. In a 1992 United Methodist Church statement on genetics we find: 'The image of God, in which humanity is created, confers both power and responsibility to use power as God does: neither by coercion nor tyranny, but by love. Failure to accept limits by rejecting or ignoring accountability to God and interdependency with the whole of creation is the essence of sin' [8]. The sin of 'failure to accept limits' is the sin of Prometheus, Frankenstein, and *Jurassic Park*. To protect the integrity of the genome is the way we would avoid such sin, so it would seem.

Is the Gene Myth Warranted?

Does the gene myth accurately reflect what scientists actually know about human genetics? No. Even though our genome certainly sets parameters and provides irreplaceable information for phenotypic development, the genes do not act alone. They are not puppeteers solely responsible for pulling the phenotypic strings. 'Genes are not puppet masters or blueprints', writes Matt Ridley [9]. The old dialectic between nature and nurture still applies.

When the HGP began, molecular biologists were estimating that the human genome would contain 100,000 genes. James Watson even put the estimate at 248,000 genes [1, p 168]. The assumption made in the 1980s and 1990s was that genes are determinative and, because the human being appears to be more complex than any other animal, it follows that humans should have more genes. Here was the assumption: the greater the complexity, the greater the number of genes. What a flabbergasting surprise, then, as HGP neared completion to find only 30,000 or so genes! Some worms have 26,000 genes and a grain of rice 50,000 genes. If the original assumption held, then a grain of rice would be more complex than Leonardo da Vinci or Werner Heisenberg. Something seemed to require revision here.

It has been speculated that what makes us distinctively human lies not in the genes themselves but rather in their protein production [10]. A single gene seems to be able to generate multiple proteins. 'Some observers have taken this as a real insult to human complexity', Collins [11] writes; 'Our complexity must arise not from the number of separate instruction packets, but from the way they are utilized. Perhaps our component parts have learned how to multitask?' Regardless of the viability of this multitask hypothesis, the human genome began to lose some of its cultural magic. The tie between the genome and human essence began to loosen. Science itself was withdrawing from considering DNA to be sacred. Yet, while the science has moved on, we in the wider culture persist in tacitly attributing sacredness to DNA.

In the recent half decade increasing attention has been given by researchers to the interaction between genes and their immediate environment, the cell. A more holistic or systemic description of cell activity inclusive of gene expression is coming to the fore. 'It has become apparent that the gene-function relationship is not so clear cut', writes Chouard [12, p 301]. To track the relationship between gene expression and phenotype we must now 'consider the entire system with its inputs, outputs and the connections in between. Where evo-devo has met systems biology, a new discipline – systems evo-devo – has emerged' [12, p 300].

When it comes to understanding the human being, the rise of neuroscience is placing considerable distance between gene expression and what actually happens in daily life. No simple straight line between gene expression and behavior can be drawn. Rather, a circuitous route through the complexities of brain structure and activity are required. 'Genes do not specify behavior directly but rather encode molecular products that build and govern the functioning of the brain through which behavior is expressed. Brain development, brain activity, and behavior all depend on both inherited and environmental influences, and there is increasing appreciation that social information can alter brain gene expression and behavior' [13, p 896]. No longer should an image of our genome as the puppeteer guide our conjectures.

Celia Deane-Drummond [14], a theologian as well as a molecular biologist, points out that within science itself we can see a withdrawal from the essentialist interpretation of the genome. 'The idea that DNA is like a predetermined "blueprint" or "code" for life does not do justice to the fluid nature of the way genes are expressed in particular contexts.... There is no need to invoke theological frameworks in order to confront the ideal of genetic determinism; it is happening, as it were, from within biological science itself' [14]. In sum, no longer does scientific justification for treating DNA as sacred exist, if it ever did.

As theologians move from what can be known scientifically about the human genome toward ethical implications, it would be advisable to focus on the human person. Our genetic makeup plays a significant role in our personhood, to be sure; yet the ethical subject should be the person and not the genetic code. Since the Second Vatican Council (1962–1965), Richard Gula [15] finds a 'noticeable shift from using the language of "human nature" to that of the "human person"'. This shift to person language will be helpful when drawing moral implications from what we know about human genetics. The human person should be the subject of ethical deliberation, not the human genome.

The Chimera Question

One area of genetic research which makes vivid the gene myth and the concern over playing God is chimerism. The chimera question elicits fears that altering the human genome will violate our human essence, that playing God will compromise the dignity of what is essentially human. The ethical question is this: should we mix the genes of two different individuals? Should we mix the genes of humans and animals?

With the term 'chimera' we refer to a single being with two or more genomes. A 'chimera' is 'a single biological entity that is composed of a mixing of materials from two or more different organisms', say Robert and Baylis [16]. Chimeras come in multiple forms: human to human or HH; human to nonhuman or HNH; and nonhuman to human NHH.

HH chimeras occur naturally. An example would be the establishment of a new embryo in a mother's body. In the case of a double fertilization and fusion, a person can be born with two genomes. He or she is a chimera. Most chimeric people grow old and die and never realize their double genetic identity. HH chimeras can also be made in the laboratory or hospital. Leukemia patients who receive blood stem cell transplants become chimeric, now carrying the genetic codes of two persons. Even in an age of genetic essentialism, to date little or no moral objection has been raised regarding HH chimerism.

What about HNH chimerism when transferring human tissue into animals? Sheep with implanted human genes now make vaccine for human diabetes patients. Such chimerism has 'not raised any outcry', comments Stanford University Law School Professor Henry Greely [17].

However, an allegedly urgent problem seems to arise with certain experiments in HNH and NHH chimerism. When human stem cells are placed in the brains of higher primates, a fear is voiced that the primates might become more humanlike. Perhaps regenerative human stem cells will elicit human thinking and self-consciousness in monkeys. If primates are almost human, might the insertion of human stem cells into monkey brains prompt them to cross a now invisible threshold into a semi-human realm? Would the line between the human and the non-human become blurred? Would this lead to moral confusion? Would our moral confusion constitute nature's backlash against our manipulating DNA? Might worse things happen?

Or, in reverse, to propose that nonhuman genetic material be placed into a human embryo elicits cries against playing God. Greely [17, p 20] continues: 'Chimeras that are produced "naturally" seem to raise few concerns. Many "unnatural" chimeras are also uncontroversial. Chimeras made by moving nonhuman parts into human beings would raise concerns when they are significant enough to cast doubt on the humanity of the recipient.... The fact that something is or isn't a chimera does not in itself raise ethical concerns'. What we see here is concern for what is distinctively human, what is essentially human. What is essentially human should be protected; and the way to protect the human genome is to prevent mixing it with nonhuman genes. So, it seems to be assumed.

In 2005 a team of scientists and ethicists writing for *Science* magazine studied the moral issues surrounding transplantation of human neural stem cells into the brains of nonhuman primates (H-NHP). Here is how they posed the questions: 'If human neural stem cells were implanted into the brains of other primates what might this do to the mind of the recipient?' And, 'could we change the capacities of the engrafted animal in a way that leads us to reexamine its moral status?' [18]. Or, to put it another way, could we change nature so that its nature would now be unnatural? And, if we could, would this be immoral?

In answer to the first question – should we mix the cells and thereby the genes of two different species? – is clear for Greene et al. [18]: 'We unanimously rejected ethical objections grounded on unnaturalness or crossing species boundaries'. Two reasons are given for making this question such an easy one to answer. First, such a thing as a boundary between species does not exist, scientifically speaking. To be sure, we have inherited from the era of Darwin and before the notion that a species can be defined as a group of organisms that can reproduce. If you cannot make babies together, then you belong to another species. But, more recent knowledge regarding the consistency of DNA shows the genetic continuity of all living things. So, thick lines of separation are fading away. 'The notion that there are fixed species boundaries is not well supported in science or philosophy. Moreover, human-nonhuman chimerism has already occurred through xenografting.' Chemically, all life belongs together. DNA continuity is what makes mixing genes easy to do.

The second question – the question of moral status – was dismissed by Greene et al. [18]. Here is the argument. To mix genes does not create a creature that is unnatural. New creatures – even new species – are just as natural as their predecessors. In addition, to object to anything a scientist does as 'unnatural' does not in itself provide a moral warrant. It is a fallacy to move from what we *describe* in nature to what we ought to *prescribe* scientists to do. This team flatly avoids giving credibility to those who commit the naturalistic fallacy.

Now, as the theologian observes what is going on here, we need to ask about the implications. One implication might be this: it would be difficult to construct a theological argument against chimerism, even HNH or NHH chimerism. According to the Bible, all creatures within God's created order share continuity of finite existence. Australian theologian Mark Worthing [19] makes this point. 'Both animals (Genesis 2:7) and humans (Genesis 1:30) are described as being "living souls". There is an inescapable boundness of human beings with the rest of God's creation. The discovery in our own time that all things living share the same building blocks of DNA serves as a further reminder of this millennia old theological truth' [19, p 91]. It appears that the Bible and Charles Darwin would agree, the difference between humans and animals is a matter of degree, not kind.

Such an observation regarding the continuity of DNA applied to primates with people could be used to justify experimental H-NHP chimerism, because the closer humans are to primates the more accurate the experimental results. The difference in the number of genes between humans and chimpanzees is only 1.5%. This suggests that chimpanzee experiments are likely to yield more valuable information than mouse experiments. This provides a scientific warrant for proceeding with primate research.

Still, one might argue quite to the contrary, the continuity observation could just as well provide an argument to prohibit mixing human and primate genomes. If primates and humans are genetically proximate, might the dignity we apply to humans drift over and apply to primates as well? Might we find we would have to apply to nonhuman animals the same rights we apply to human persons? This would create social and cultural confusion. Could we tolerate such confusion? Even if such confusion might become a cultural problem, would it also become a theological problem? The theological discussion is not yet mature enough to forecast where the theological community will come down. Be that as it may, it appears to me that no theological warrant for protecting the sacred character of the existing human genome from contamination by an animal genome exists [20].

Avoiding Bio-Idolatry

Theologians in the Reformed tradition are particularly alert to false sacreds. They are the first to blow the whistle to warn of approaching threats against the First Commandment, the commandment against substituting an idol for the true God. According to John Calvin, God the creator does not share divinity with what is created. 'Nothing belonging to his divinity is to be transferred to another' [21, p 1:117]. God alone is sacred. Nothing within the created order – even DNA – ought to be deemed sacred. Yet, we human beings have a proclivity for attributing sacredness to non-sacred things. To paint the created order with sacred colors constitutes idolatry, according to Calvin. 'Man's nature, so to speak, is a perpetual factory of idols' [21, p 1:108]. Perhaps, we might surmise, in our own era DNA has become a candidate for false sacredness – that is, for idolatry.

Might treating DNA as sacred risk idolatry? Bouma et al. [22] follow Calvin in positing that 'God is the creator. Therefore, nothing that God made is god, and all that God made is good.' They proceed to draw out implications here for the confusion created by the commandment against playing God. We must avoid idolatrous expectations from our technology, to be sure; 'but to presume that human technological intervention violates God's rule is to worship Mother Nature, not the creator. Natural processes are not sacrosanct' [22]. What this clarifies is that what is natural – and our DNA is natural – belongs to the created order of things, not inherent sacred quality which requires that we keep our hands off of it. Human intervention will not violate its divinity, nor will it sacrifice human dignity. In short, genetic science and interventionist technology do not profane the human genome.

If we cannot find an inviolable human essence in our genome, then where should Christians go to look for it? According to Karl Barth [23], we look for the essence of what makes us human not to DNA – not to anything in our physical nature alone – but rather to Christ. Christ, the new Adam, marks what is essential to human nature. And, according to Barth, this Christological anthropology includes the two natures. It includes the divine as well as our naturally conceived human nature. Christ's incarnate divinity, wherein the human will is in harmony with the divine will, marks the essence of human nature. 'His [Christ's] essence is that of a man like ourselves, the individual soul of an individual body, knowing and willing and feeling as a man, active and passive in the time allotted, responsible to God and tied to his fellows. What the grace of His origin does involve and effect, with supreme necessity and power, is the exaltation of His human essence. Exaltation to what? To that harmony with the divine will' [23, IV/2:91–92]. When looking for the essence of what makes us human, we should look not at our genome but rather at the person of Jesus Christ and to the harmony of will he enjoyed with God.

From Nature to Naturalism

The question of the sacredness of the human genome arises because of some slippage between science understood as a method for pursuing empirical research and science understood as an ideology, as an '…ism'. In this case, we are looking at science as a form of 'naturalism'. 'In the broadest sense, *naturalism* can denote any philosophy in which "nature" or "the natural" functions as the most general explanatory or normative concept' [24]. In naturalism, nature has become much more than merely the subject matter that scientists study. Nature has surreptitiously become identified with 'essence'. Nature has become subtly sublime, because it is thought to provide the fundamental characteristics of an entity. And when nature is thought be equivalent to essence, it takes on moral valence. Nature becomes the source of value and the reference point for moral normativity.

When nature – or something within nature – becomes sacred, the scientist becomes the high priest who provides access to the sacred while protecting it from being profaned. Science replaces the prophets and seers of religion, says astrophysicist Adam Frank [25], because science provides a much more advanced revelation of the sacred. Like yesterday's myths, today's science listens to the revelation of the sacred within our material world. 'The constant fire is the aspiration to know what is essential, what is real, what is true. It emerges from the elemental experience of the world as sacred. Mythic narratives are one expression of that aspiration. Scientific narratives are another … science retains living roots in myth' [25]. In short, science today plays the role that religion allegedly played in pre-modern eras. At least, this is the doctrine touted within our culture by the scientific priestly caste.

Today's naturalism need not be quarantined in the laboratory. No fence separates naturalism from survivals of traditional religion. They can overlap. Church goers and members of a variety of religious communities feel they can embrace some aspects of the naturalist position. The logic goes like this: if the natural world is God's creation, then perhaps we can hear God's voice speaking to us from within the natural order. In our culture, naturalism takes the form of an intuition, a vague sense that nature provides moral guidance. In theological schemes which incorporate 'natural law', nature is the voice of God. 'The natural law approach is of lasting value for today in that it grounds an experiential morality while holding to an ideal of shared human truth, and manifests a confidence that God's will for persons is revealed in creation as an ongoing process of discovering God in human life' [26, p 215]. What this means for one of America's most respected Roman Catholic bioethicists, Lisa Sowle Cahill [26], is that nature understood by science and natural law theology may share a foundational role in ethical deliberation.

Yet, we must ask: does it follow from fundamental theological commitments that one aspect of nature – human DNA – should be deemed sacred? No.

I recently served on a research project at Rice University funded by the Ford Foundation to deal with the ethical implications of 'altering nature'. We solicited contributions from experts on religious traditions such as Judaism, Islam, Roman Catholicism, Protestantism, Hinduism, Buddhism, as well as feminist theology. We examined carefully the cosmologies and anthropologies of each, looking for the level of commitment or non-commitment to a naturalistic point of view. We specifically looked for religious grounds for deeming the human genome sacred. No such religious justification could be found. 'As we have interrogated the fundamental anthropologies and understandings of nature in the world's religious traditions, we find no prior commitments that would lead contemporary religious adherents to think of DNA as sacred. … This leads to a principle: *DNA is not sacred*' [27, p 156].

References

1 Kevles D, Hood L (eds): The Code of Codes: Scientific and Social Issues in the Human Genome Project. Cambridge, Harvard University Press, 1992.
2 Jaroff L: The Gene Hunt. Time, March 20, 1989, vol 133, No 11, p 67.
3 Nelkin D, Lindee MS: The DNA Mystique: The Gene as a Cultural Icon. New York, Freeman, 1995.
4 Maddox J: Has nature overwhelmed nurture? Nature 1993;366:107.
5 Hubbard R, Wald E: Exploding the Gene Myth. Boston, Beacon Press, 1993.
6 Peters T: Playing God? Genetic Determinism and Human Freedom. London, Routledge, 2002.
7 Kass L: Life, Liberty and the Defense of Dignity: the Challenge of Bioethics. San Francisco, Encounter Books, 2002, p 149.
8 United Methodist Church Genetic Task Force: Report to the 1992 General Conference. UMC, 1992, p 114.
9 Ridley M: Nature via Nurture: Genes, Experience, and What Makes Us Human, San Francisco, Harper, 2003, p 6.
10 Lander FS, Linton LM, Birren B, et al: Initial sequencing and analysis of the human genome. Nature 2001;409:860–921.
11 Collins F: The Language of God. New York, Free Press, 2006, p 125.
12 Chouard T: Beneath the surface. Nature 2008;456: 300–301.
13 Robinson G, Fernald RD, Clayton DF: Genes and behavior. Science 2008;322:896–898.
14 Deane-Drummond CD (ed): Brave New World? Theology, Ethics and the Human Genome. London, T.&T. Clark, 2003, p 227.
15 Gula RM: Reason Informed by Faith. New York, Paulist Press, 1989, p 63.
16 Robert JS, Baylis F: Crossing species boundaries. Am J Bioethics 2003;3:1–13.
17 Greely H: Defining chimeras and chimeric concerns. Am J Bioethics 2003;32:17–20.
18 Greene M, Schill K, Takahashi S, et al: Moral issues of human-nonhuman primate neural grafting. Science 2005;309:385–386.
19 Worthing MW: Human and animal intelligence: a difference of degree or kind?' in Kelly TJ, Regan HD (eds): God, Life, Intelligence and the Universe. Hindmarsh, Australian Theological Forum, 2002, pp 85–110.
20 Peters T: A theological argument for chimeras. Nature Reports Stem Cells. http://www.nature.com/stemcells/2007/0706/070614/full/stemcells.2007.31.html.
21 Calvin J: Institutes of the Christian Religion (1559); in McNeill JT (ed): Library of Christian Classics XX, XXI. Louisville, Westminster John Knox Press, 1960.
22 Bouma H, Diekema D, Langerak E, Rottman T: Christian Faith, Health, and Medical Practice. Grand Rapids, William B. Eerdmans, 1989, pp 4–5.
23 Barth K: Church Dogmatics. Edinburgh, T.&T. Clark, 1936–1962.
24 Stout J: Naturalism; in Jones L (ed): Encyclopedia of Religion, ed 2. New York, MacMillan, 2005, vol 9, p 6428.
25 Frank A: The Constant Fire: Beyond the Science vs. Religion Debate. Berkeley, University of California Press, 2009, p 224.
26 Cahill LS: Feminism and Christian ethics; in Mowry LaCugna C (ed): Freeing Theology: The Essentials of Theology in Feminist Perspective. San Francisco, Harper, 1993, pp 211–234.
27 Peters T, Aguilar-Cordova E, Crawford C, Lebacqz K: Religious Traditions and genetic enhancement; in Lustig BA, Brody BA, McKenny GP (eds): Altering Nature. Vol 2: Religion, Biotechnology, and Public Policy. Dordrecht, Springer Science Business Media, 2008, pp 109–160.

Ted Peters
Pacific Lutheran Theological Seminary
2770 Marin Avenue
Berkeley, CA 94708 (USA)
Tel. +1 510 524 5264, Fax +1 510 524 2409, E-Mail tpeters2ct@aol.com

Pfleiderer G, Brahier G, Lindpaintner K (eds): GenEthics and Religion. Basel, Karger, 2010, pp 118–128

The Jewish Perspective on GenEthics

Ronald M. Green

Department of Religion, Dartmouth College, Hanover, N.H., USA

Any examination of the relationship between Jewish thinking and genetics must begin with a striking paradox. No single religious group has been more greatly victimized by genetics than Jews. Under the Nazi regime, a wicked, misguided quest for genetic racial 'purity' led to the sterilization and slaughter of millions of Jews. Yet, at the same time, probably no single religious or ethnic community has been more receptive to genetic medicine than Jews. In Israel and the United States, Jews have supported major programs of genetic screening and testing in the effort to reduce the incidence of genetic diseases known to be prevalent in their communities. Among nations, Israel has the highest per capita rate of use of prenatal testing. Around the world, Jews have also been supportive of a host of genetic reproductive technologies such as in vitro fertilization (IVF), preimplantation genetic diagnosis (PGD), stem cell and even cloning research [1].

It is tempting to explain this receptivity to genetic information and technologies as a consequence of the high degree of modernization within Jewish communities and the reduced hold of traditional religious teachings. However, while differences in respect to religious authority explain some important differences among Jews in what counts as acceptable means for the use of genetic information, they do not explain the nearly universal Jewish openness to genetic information and genetic science and medicine. One sign of this is the Dor Yeshorim program [2]. Begun in the 1980s as a response to the prevalence of Tay-Sachs disease, this program now extends to carrier screening for a host of autosomal recessive disorders that are prevalent in Ashkenazi Jewish communities [2]. Dor Yeshorim has been enormously successful everywhere, reducing the incidence of Tay-Sachs by 90%. No children with Tay-Sachs disease have been born to Ashkenazi Jews who have participated in the Dor Yeshorim screening program. Some of its greatest successes have been registered in ultra-orthodox communities in New York and Israel, where arranged marriages facilitate premarital screening. In the words of Dr. Fred Rosner, a leading scholar of Jewish bioethics, 'The ultra-orthodox Jewish community will never allow a marriage to take place without being screened'. In this community 'the first question asked is, "Have you been tested?"' [3].

In what follows, I propose to explore the classical biblical and Talmudic roots of this Jewish openness to genetics. By 'genetics' here, I mean the range of medical genetic technologies. This includes preconceptional testing and screening for genetic disorders, prenatal screening and testing, including the use of PGD, chorionic villus sampling and amniocentesis, newborn screening, and adult genetic screening and testing. It also looks down the line to the combined use genetics and reproductive technologies (what some call 'reprogenetic medicine') for purposes of human genetic engineering, including disease treatment and trait enhancement. Finally, I draw some implications of Jewish teaching for non-human genetic modification, including the creation of genetically modified organisms for food or other purposes.

By 'Jewish teachings', I refer almost entirely to the corpus of Talmudic and later rabbinic writings. This corpus, with its foundation in the biblical commandments but vastly expanded by centuries of interpretation, constitutes *halacha* or Jewish law. For Torah-obedient Jews in the orthodox and ultra-orthodox communities, *halacha* is the guiding basis of their lives. Less Torah-obedient Jews belonging to the conservative, reform or reconstructionist movements tend to heed *halacha* much more selectively, finding themselves instructed by certain teachings while ignoring or rejecting others. However almost all Jews, including many secular Jews who do not otherwise pay attention to *halacha*, have been informed by its ethical sensibilities. I do not see it as misleading to speak of a 'Jewish outlook' on genetic matters. It is the *halachic* roots of this outlook that I propose to explore.

From this body of literature, three broad principles emerge. Together, these constitute the supporting pillars of Jewish biomedical ethics as it pertains to genetic science and medicine. The first of these principles or pillars is the stringent obligation to preserve human life through medical healing and medical research. We will see that this extends to eugenic measures to prevent the conception or birth of individuals with serious genetic diseases, although Jewish teachers have different views on the precise *means* by which this may be accomplished. The second pillar is what might be called a 'developmental' view that accords nascent human life greater moral weight with increasing gestational age. This 'developmental' view has shaped Jewish thinking about IVF and PGD, as well as about the permissibility of human embryonic stem cell research. The third pillar is a Jewish openness to technological interventions in nature, including human biology, for the improvement of human life. At its roots, Judaism is a tradition that sees human beings as 'co-creators' with God in creation and that does not exhibit a fear that human beings will use technology to 'play God'.

Preserving Human Life

One of the preeminent principles of Jewish ethics is the obligation to protect human life and promote health in all one's deeds, and by medical means if necessary. This is the solemn obligation of *Pikuach Nefesh*, the duty to save a human soul. Early rabbis looked to the biblical text 'And he [an attacker] shall surely cause him [his victim] to be healed', [Exodus 21:19] to conclude that 'From here we see that the physician is given permission to heal'.

Normative Judaism did not see this obligation as being in conflict with faith in God's sovereignty (Karaite exception). In religions that affirm God's mastery of the world, there is sometimes a tendency to adopt a passive or fatalistic attitude to misfortunes. Biblical texts describing God as the 'healer' [e.g. Jeremiah 17:14] might have reinforced the tendency to interpret health and sickness as entirely in God's hands. A sectarian movement within Judaism, the Karaites, actually adopted this stance, holding that 'God alone should be sought as physician, and no human medicine should be resorted to' [4]. Nevertheless, the Karaites were an isolated exception. Normative rabbinic Judaism always affirmed in the strongest terms human responsibility for the promotion of health and the prevention or cure of disease. The firm rabbinic teaching over the centuries was that, while in biblical times God may have directly caused and removed disease, 'nowadays one must not rely on miracles' according to the 18th century rabbi Chayim Azulai [5]. Medical care was always valued, and well-known rabbis like Maimonides often had distinguished medical careers. While God might be the ultimate healer, He worked His will through human agency [6]. If necessary, Sabbath restrictions must be set aside to provide lifesaving medical care [7], and the rabbi of a community who did not make this known was regarded as derelict in his duty [8]. The sanctity of life extends to the self: Judaism opposes personal recklessness and self-neglect if it endangers one's life.

A ready acceptance of genetic medical interventions, where they are available, follows from this deep basic commitment. This perspective is amplified by another Jewish obligation: the duty to fulfill the biblical commandment 'Be fruitful and fill the world' *(p'ru ur' vu)*. Understood by the school of Rabbi Hillel as the duty to have one boy and one girl, and by the school of Shammai as having two boys, this commandment underlies a strong Jewish commitment to marriage and parenthood [9–11]. In a context where health and offspring are intensely valued, reproductive and genetic medicine have both been enthusiastically received. The Talmud even supports a eugenic perspective when it counsels against one's marrying into a family of epileptics or lepers [12]. These values help explain the ready acceptance of genetic medicine by Jews around the world and in Israel.

Development and Nascent Human Life

Neither an emphasis on the sanctity of human life nor support for medical interventions adequately explains Jewish support for genetics. Other pro-natalist, medically inclined traditions such as Roman Catholicism do not exhibit quite the same degree of enthusiasm for genetics. To understand Jewish attitudes, therefore, we must invoke the two other pillars of Jewish thought I have mentioned.

The first of these is what I have termed a 'developmental' view of nascent human life. In contrast to Roman Catholicism, which accords the human embryo from conception onward a moral status equivalent to any human being, classical Jewish teaching makes a series of graded moral distinctions. Before 40 days gestational age, the embryo is regarded as having little or no moral weight. Talmudic texts characterize it as 'mere fluid' or 'like water'[13]. Because of the high value placed on procreation within the Jewish tradition, a pregnancy is valued from its very start and may not be terminated for trivial reasons. But this concern does not extend to the embryo ex utero, in a Petrie dish or freezer.

This relative devaluation of the early ex utero embryo justifies procedures such as IVF, PGD, and human embryonic stem cell research in which embryos are manipulated or destroyed whether for the relief of infertility or other medically compelling reasons. Israel has more infertility clinics per capita than any nation in the world. As part of the basket of health services, every Israeli citizen is entitled to unlimited rounds of IVF treatment, up to the birth of two live children. Israel's 300,000 ultra-orthodox Jews, roughly 5% of the population, fully avail themselves of this and have established special programs for rabbinic surveillance of IVF procedures [1]. Israel is also a world center of embryonic stem cell research. In the United States, the leading reform, conservative, and orthodox Jewish communities have all given their ethical endorsement to embryonic stem cell research, seeing it as not only permissible under Jewish law but as an embodiment of Jewish values.

Recently, some Jewish bioethicists have questioned whether the classical Jewish devaluation of the very early embryo any longer makes sense in the light of modern biological knowledge. Arguing for a greater degree of rapprochement between Catholic and Jewish bioethics, for example, Cohen [14] has asked whether it is morally responsible for Jews to seek the fruits of modern science in the form of stem cell research 'without confronting the facts of modem embryology in order to understand what embryos really are'. David Novak, a leading Jewish ethicist, has also asked whether the Talmudic devaluation of the embryo can withstand modern scientific scrutiny, especially the understanding that even the earliest embryo possesses a complete and self-developing genome. Novak took this position during a text analysis session at the annual meeting of the Society of Jewish Ethics (January 6, 2007).

While there is some merit to these arguments, they fail to consider that rabbinic teaching in this area may be well grounded both in medical reality and a non-genetic view of what constitutes human moral worth. With the active involvement of some of their number in medical care, including surveillance of early spontaneous miscarriages, rabbinic decisors were familiar with the unformed and amorphous nature of the early embryo. It was this entity, lacking organs or bodily form and the physical basis for either perception or cognition, that they judged to be of negligible moral worth. In simple terms, they were never persuaded that such an entity, despite its potential, merits significant moral regard. This estimate might well be impervious to the kinds of refined embryological knowledge alluded to by Cohen or Novak.

This graduated and developmental view of nascent human life continues in rabbinic teaching as the embryo develops into a fetus. Here, guided by key biblical texts such as Exodus 21: 22–25 where the death of a fetus resulting from its mother's injury is not treated as a capital offense, the rabbis consistently affirmed that until the fetus emerges from the birth canal it is not to be considered a *'nefesh'* or living soul for moral or juridical purposes [15].

Among other things, this led to the teaching that abortion is required when the continuation of a pregnancy places the mother's life or health in jeopardy. This view finds expression in one of the core texts of the Talmud, the *Mishnah* (or pre-Talmudic, recorded oral teaching) of the Talmudic Tractate Oholoth:

> 'If a woman is in hard travail, we cut up the child in her womb and brings it forth member by member, because her life comes before that of the child. But if the greater part had proceeded forth, one may not touch it, for one may not set aside one person's life for that of another' (7:6).

In striking contrast to most Christian traditions, abortion in these cases is neither morally prohibited nor morally permitted: it is required. The overriding obligation of *Pikuach Nefesh* mandates saving the mother and destroying the fetus, since the latter is not regarded as a living soul.

Although the Oholoth text is unambiguous about abortion where the mother's life is in peril, it is unclear with regard to the permissibility of abortion for lesser reasons. This has occasioned a longstanding debate within rabbinic sources marked by two competing positions. On one side are those who refuse to permit abortion for lesser reasons than saving the life of the mother, including perhaps grave genetic defects in the fetus. Genetic abortion is allowed only when the birth of a genetically deformed child might imperil the mother's life, as when she has likely to commit suicide as a result. This position closely approximates stringent Christian views, although formal Catholic teachings rule out abortion even to save the mother's life. On the other side are those who would permit abortion for lesser reasons, including severe congenital anomalies and any degree of maternal distress at the possibility of giving birth to a genetically impaired child.

The stringent view relies heavily on a gloss on this passage by Maimonides [16]. In explaining the teaching in Oholoth, Maimonides likens the fetus to a pursuer *(rodef)* who has placed the mother's life in jeopardy. Maimonides' meaning, some decisors conclude, is that the fetus has the same status as an adult 'pursuer' whose life may be taken because it threatens one's own. The implication is that the fetus, like the adult 'pursuer', may not be killed for lesser reasons.

This stringent viewpoint has a bearing on the permissibility of genetic screening and testing programs. Screening programs are permissible if they prevent the marriage of individuals carrying a genetic disease. This has been the premise of the Dor Yeshorim program which, as I said, has largely succeeded because of the prevalence of arranged marriages in ultra-orthodox communities. In these communities all young people are genetically screened and the results are maintained in a confidential registry which is consulted only when a marital union is proposed. But most orthodox and ultra-orthodox rabbis have not supported programs of prenatal testing for genetic diseases since this is likely to lead to the abortion of a defective fetus. This concern informed the thinking of Rabbi Immanuel Jakobovits who served as the Chief Rabbi of Great Britain and wrote widely on Jewish biomedical topics. Jakobovits [17] further based his opinion on what he regarded as the uncertainty of any genetic diagnosis in utero and the fact that the aborted fetus might not be seriously diseased. It is to be noted that despite his affirmation that the only legitimate indication for an abortion is a threat to the mother, Rabbi Jakobovits has repeatedly affirmed that every individual case of possible genetic deformity must be judged by competent rabbinical experts [12, p 263]. A similar position has been taken by Rabbi I.J. Unterman, former Chief Rabbi of Israel, who, in explaining his position, placed even greater stress on the sanctity of prenatal life. In a discussion of fetuses possibly damaged by exposure to rubella or thalidomide, Unterman states that 'this prohibition would stand even if it were certain … that the child would be born deformed, just as it is forbidden to kill a crippled person' [17].

In contrast, the lenient Jewish view on abortion has as its point of departure in a different Talmudic text, one found in the Tractate 'Arakin (7a). It deals with the case of a woman who has already been convicted of a capital offense and who is discovered to be pregnant. The question is, may the execution be delayed – even for one day if need be – to spare the life of the child she is carrying? 'Arakin's answer is blunt:

> 'If a woman is about to be executed, one does not wait for her until she gives birth: But if she had already sat on the birthstool, one waits for her until she gives birth.'

In the Gemara or commentary that follows this Mishnah, the assumed reason for the ruling is given:

> 'It is necessary to teach it for one might have assumed since Scripture says: According as the woman's husband shall lay upon him (Exodus 21:22), that it [the unborn child] is the husband's property, of which he should not be deprived, therefore we are informed [that it is not so].'

This is a dramatic text. Superficially, it seems ghastly, which may be why even modern reform or conservative defenders of the right to abortion rarely cite it as evidence of Jewish thinking on abortion. Nevertheless, within a context where capital punishment is assumed, what is noteworthy is the fact that the major concern in this passage is directed toward the suffering experienced by the woman. As commentators note, she is naturally in anguish about her impending execution. Hence no delay is permitted if this prolongs her suffering by even a few hours. The concern for her is evidenced as the Gemara on this text continues:

> 'What is the reason [for this teaching]? – As soon as it moves (from its place in the womb) it is another body, Rab Judah said in the name of Samuel: If a woman is about to be executed one strikes her against her womb so that the child may die first, to avoid her being disgraced [by her bleeding should a living child emerge].'

The import of these passages is that the mother's feelings take priority over the fetus's life and over the father's property rights in the fetus. The prospect of extending her suffering or of worsening it by adding an element of physical shame suffices to justify feticide. Once the child has substantively passed through the birth process this permission ends, but full legal sanctity does not attach to the preterm child.

Clearly this text is not an eccentricity. It is in direct continuity with the *halakhic* interpretations of Exodus 21:22–25 and with some leading commentators' interpretation of our previous text from Oholoth. Throughout all these texts, the conclusion is the same: the fetus is not a 'life', a *nefesh*; in the eyes of Jewish law it is not a person. Prior to birth, therefore, any significant threat to the mother's physical or emotional wellbeing justifies the termination of a pregnancy.

The 'Arakin text underlies a lenient series of more contemporary teaching about genetic disease and genetic medicine. Most of these views base themselves on a ruling by the 18th century Rabbi Jacob Emden. The question before Emden was whether an adulteress might hide her crime by aborting her bastard child. Part of Emden's response was based on an ingenious application of the 'Arakin text. Since the mother would be put to death if tried and convicted of her crime, said Emden, and since Jewish Law allows killing the fetus in such cases, this child may be regarded as dead and its life may be taken to save the mother's! But Emden also went beyond this. He ruled that in view of the reduced valuation of the fetus in Jewish law, abortion is permissible in any case of 'great need' on the mother's part. Her life, he argued, did not have to be in danger [18, 19].

In our own day, Emden's position has been expanded by some commentators. One writer, for example, has extended it to permit abortion for even a 'slim reason' on the mother's part [19]. And this same ruling has become a major precedent in the lenient approach to the question of aborting a fetus with Tay-Sachs disease. In an important decision published in 1976, Rabbi Eliezer Waldenberg, former head of the court of Jewish Law in Jerusalem, ruled that it is permitted to abort a Tay-Sachs fetus until the 7th month of pregnancy. He later extended this permission to any other case of 'great need' including the case of a deformed Down's syndrome fetus and fetuses with other genetic conditions that would place the family or the mother under great stress [18, 19]. Other lenient commentators have arrived at Rabbi Waldenberg's conclusion by a slightly different route. They have ruled that since the severely deformed fetus is not a viable life, aborting it in late pregnancy is not culpable under Jewish law [20].

Whether on the right or the left of *halachic* opinion, the Jewish legal tradition is open to efforts to use genetic medicine to avoid the birth of children with serious genetic disorders. Premarital screening is universally accepted, while PGD achieves nearly the same level of approval. On a case-by-case basis, even late-term genetic abortion for very severe disorders will also be approved by many orthodox rabbinic authorities. Human embryonic stem cell research aimed at clinical therapies or to understand the etiology genetic diseases is also widely supported by Jewish teachings.

Openness to Technology

The third supporting pillar of Jewish thinking about genetic science and medicine, and again one that distinguishes it from some Christian and other religious traditions, is strong support for the advancement of biomedical technologies and an encouragement to human interventions in nature. From its inception, the Jewish tradition has not feared human beings 'playing God' in relationship to the natural world, and has even regarded them as 'co-creators' with God in improving the natural order. Jewish writers frequently invoke Psalm 115: 6 in this connection: 'The heavens are the heavens of God' yet 'the earth has He given to the sons of man'.

Particularly noteworthy are rabbinic interpretations of the statement 'I, the Lord Shaddai' found in Genesis 17:1. In rabbinic exegesis the divine appellation 'Shaddai' is understood as an acronym for the phrase 'Who said to My universe, "Enough!"' As David Bleich [21] observes in his survey of *halachic* literature bearing on genetic engineering, the word 'enough' here is taken by various commentators to mean that God, when He created various natural entities, arrested their development before completion. Thus, wheat plants produce the seed kernel, but not bread. As Bleich says,

'The Creator could well have made it possible for the kernels to crumble into flour, for the flour to absorb rain or moisture from the atmosphere, for the wind to churn the water-drenched flour so that dough be formed and for the heat of the sun to bake the mixture in order to yield a product that might literally be termed a "breadfruit". Instead, the Creator arrested the process long before its completion and ordained that grinding the wheat, mixing the flour with water, kneading the dough and baking the bread be performed by man.'Bleich continues:

> '... in the end of days, God will allow the processes of creation to reach their destined end by modifying the natural order in a manner that will permit the creative process to become complete and thus spare man any travail. In the interim, however, He has declared, "Enough!", i.e., He has precipitously interrupted the process of creation and co-opted man, who must complete the process, as a collaborator in fashioning the universe' [21].

In Bleich's view, human beings' role in the 'completion' *(hashlamah)* of the process of Creation extends to the genetic manipulation of other species and the creation of novel organisms and foodstuffs by genetic means. Bleich takes note of several specific biblical commandments [Leviticus 19:19] that prohibit the intermixing of diverse species of plants and animals, but he observes that in the commentary tradition, the scope of this prohibition has been interpreted to exclude only the sexual crossbreeding of animal species and certain kinds of plant grafting, but not cross-pollination procedures. Furthermore, these prohibitions pertain only to Jews and are not regarded as of universal import. For Bleich the genetic engineering of plants and animals is fully within the scope of permissible human activities as an expression of our role as co-creators. Other commentators arrive at a similar conclusion by observing that Talmudic prohibitions do not extend to the microscopic level [22].

Does this co-creatorship extend to the alteration of our own genetic material? Where disease conditions are involved, Jewish teaching would permit both somatic and germline gene therapies. Indeed, the obligation of *Pikuach Nefesh* makes such therapies mandatory. In the words of Fred Rosner 'genetic manipulation is not considered to be a violation of God's natural law but a legitimate implementation of the biblical mandate to heal' [3]. Invoking the commandment *p'ru ur' vu*, some orthodox Jewish commentators have even supported human cloning as an alternative for individuals otherwise unable to reproduce (see the testimony of Rabbis Moshe Tendler and Elliot N. Dorff before the National Bioethics Advisory Committee, March 14,1997) [23]. However, it is also true that the principle of *Pikuach Nefesh* militates against the utilization of life-threatening and untested experimental genetic therapies. This is true of human manipulations, and by extension, interventions in the natural order that imperil human life. Although Judaism does not invoke a 'precautionary' principle in this connection, human safety and wellbeing is always a first consideration in relation to the introduction of novel technologies.

One final question. If Jewish teaching both permits and requires the implementation of safe and effective genetic therapies that are aimed at preventing or curing disease, does it also permit the use of genetic enhancements, those designed to make a person 'better than well'? One orthodox commentator, Michael J. Broyd, believes so:

> 'Using PGD to create a child without a specific illness would seem to be permitted according to Jewish law at the discretion of the child's parents. The same can be said for PGD that is designed to enhance any given characteristic in a child that increases a child's ability or functionality, in the discretion of the parents. However, were PGD used in a manner that were designed to harm the child, even if the parents sincerely believed that the conduct they were embarking on was not harmful, such activity would seem to violate Jewish law and in this circumstance would empower a Jewish court (as well as a secular court) to intervene to prevent such conduct. When it comes to the rights of parents, Jewish law only gives parents the right to conduct themselves in the best interests of their children, bounded by the range of activities found in the best interest of the child' [24, 25].

Conclusion

In my recent book, *Babies by Design* [26], I have argued for the permissibility of human genetic self-modification, including forms of human enhancement when these prove themselves to be safe and effective. Although I am an ethicist of Jewish background, I did not make explicit appeal to my own religious tradition in making that argument. In reexamining the Jewish materials, however, I realize how much I have been informed by my own tradition. Jews everywhere, whether devoutly obedient to Torah or highly secularized, share what I have elsewhere called a 'bioethical sensibility', a similar resonance to the ethical issues that arise with biomedical technologies [27]. By identifying these pillars of Jewish thought, I have tried to illustrate the contribution of this tradition to contemporary thinking about genetics.

Jewish teachings share common ground with Christian, Muslim and other traditions, especially when it comes to the validity of therapeutic genetic interventions.

But they differ from conservative Christian teachings when these interventions implicate human embryos. There is also a different attitude or tone in Jewish thought, less concern with human beings usurping God's role in creation and more emphasis on their role as God's active partners in improving creation. Indeed, one orthodox Jewish commentator finds warrant for this in circumcision, the distinguishing ritual sign of males' membership in the Jewish community. Circumcision, Rabbi Moshe Tendler has observed, is the first post-Creation enhancement (remarks delivered at a conference on Jewish Biomedical Ethics: Comparison with Christian Perspectives, Georgetown University, Washington, April 1996).

Debates about genetics will occupy us in the decades ahead. In these debates, an awareness of the distinctive nature of Jewish teachings can serve two purposes. First, it can help broaden our understanding of the range of possible moral approaches to these issues. Second, it can warn religious individuals of any tradition against believing that their very strong views are the only spiritually informed way of looking at these matters. The distinctive teachings of Judaism remind us that a tradition close to the source of all the biblical faiths has sometimes drawn other conclusions from our common heritage.

References

1 Kahn SM: Rabbis and Reproduction: The Uses of New Reproductive Technologies among Ultraorthodox Jews in Israel. Hadassah International Research Institute on Jewish Women, Brandeis University, Working Paper No. 3, August 1998. http://www.brandeis.edu/hbi/pubs/Kahnworkingpaper.pdf.

2 Dor Yeshorim: http://en.wikipedia.org/wiki/Dor_Yeshorim.

3 Rosen C: Eugenics – Sacred and Profane. The New Atlantis 2003, No. 2, pp 79–89.

4 Kaufmann K, de Harkavy A: Karaites and Karaism. Jewish Encyclopedia. http://www.jewishencyclopedia.com/view.jsp?artid = 108&letter = K.

5 Rosner F: Modern Medicine and Jewish Law. New York, Bloch, 1972, p 24.

6 Babylonian Talmud, Tractate Baba Kamma, 85a.

7 Babylonian Talmud, Tractate Shabbath, 150a; Tractate Yoma 84b.

8 Yoreh De'ah 336:1; in Bleich JD: Judaism and Healing. New York, Ktav, 1981, p 4.

9 Gordis R: Be fruitful and multiply – biography of a Mitzvah. Midstream 1982;28:21–29.

10 Gordis R, Jakobovits I: Large Jewish families. Midstream 1983;29:39–43.

11 Gordis R: Love and Sex: A Modern Jewish Perspective. New York, Farrar Straus Giroux, 1978, Ch 8.

12 Jakobovits I: Jewish Medical Ethics. New York, Bloch, 1975, p 155f.

13 Tractate Yebamoth, Babylonian Talmud, 69b, Tractate Niddin, Babylonian Talmud, 3, 7, 30b.

14 Cohen E: A Jewish-Catholic Bioethics. First Things June/July, 2005, pp 7–10 at 7.

15 Babylonian Talmud, Tractate Sanhedrin, 72b.

16 Maimonides M: The Code of Maimonides (Mishneh Torah). New Haven, Yale University Press, 1949, vol 11, p 196.

17 Jakobovits I: Review of Recent Halakhic Periodical Literature: Deformed Babies. Tradition 1963, vol 5, No. 2, p 268.

18 Sinclair D: The Legal Basis for the Prohibition on Abortion in Jewish Law. Israel Law Review 1980, vol 15, pp 124, 126f.

19 Feldman DM: Marital Relations, Birth Control and Abortion in Jewish Law. New York, Schocken Books, 1974, pp 288f, 291–294.

20 Jakobovits I: Tay-Sachs disease and the Jewish community. Proc Assoc Orthodox Jewish Scient 1979;5:11–17.

21 Bleich JD: Survey of recent Halakhic periodical literature: genetic engineering. Tradition 2003;37:66–87.

22 Rosenfeld A: Judaism and gene design; in Rosner F, Bleich JD (eds): Jewish Bioethics. New York, Sanhedrin Press, 1979, pp 401–408.

23 Broyde MJ: Cloning People and Jewish Law: A Preliminary Analysis. Jewish Law Articles. www.jlaw.com/Articles/cloning.html.
24 Broyde MJ: Pre-implantation genetic diagnosis: stem cells and Jewish law. Tradition 2004;38:54–75 at 65.
25 Broyde MJ: Child custody and Jewish law: a review. J Halacha Contemp Soc 1999;36:21–46.
26 Green RM: Babies by Design. New Haven, Yale University Press, 2007.
27 Green RM: Religion and bioethics; in Davis DS, Zoloth-Dorfman L (eds): Notes from a Narrow Ridge. Frederick, University Publishing Group, 1999, pp 165–181.

Ronald M. Green
Department of Religion, Dartmouth College
27 North Main Street
Hanover, NH 03755 (USA)
Tel. +1 603 646 1264, Fax +1 603 646 9393, E-Mail ronald.m.green@dartmouth.edu

Pfleiderer G, Brahier G, Lindpaintner K (eds): GenEthics and Religion. Basel, Karger, 2010, pp 129–137

Human Genetic Technologies and Islamic Bioethics

Siti Nurani Mohamed Nor

Department of Science and Technology Studies, University of Malaya, Kuala Lumpur, Malaysia

Asia is the hub of biotechnological superpowers – South Korea, Japan, India and China are receiving enormous federal backing and funding to intensify research in biotechnology. Muslim societies have advanced tremendously due to modern cultural standards, high standards of education and healthcare, and the surge and influx of information through the internet and cable TV. Universities are competing with each other to expand their research programs, determined not to be left out of this biotechnological century.

Egypt, which houses a notable institution like the University of Al-Azhar which other Muslim countries look up to for *fatwas* or Islamic rulings on contentious issues in the life and medical sciences, is itself initiating research using embryonic stem cells left over from infertility treatment [1]. Embryonic stem cells possess pluripotent qualities which means they are capable of issuing into a number of cell types. Research on such entities has enormous benefits and appears to promise an end to human suffering. The Islamic precept, 'to avoid evil, harm or sufferance and to seek benefits' seems to favor therapeutic cloning, stem cell and gene therapy. Muslim scientists argue that early embryos less that 14 days old have 'no life' and, rather than leave them to perish, why not use them for research for the benefit of human beings? There is an air of general acceptance for stem cell research in countries such as Iran, Saudi Arabia, Singapore and Malaysia, in the midst of debates and pending approvals for such research.

Muslim legislators begin to fear that the New Human Sciences would open up Pandora's box and unleash the unknown. How can they wisely attend to the myriad of ethical problems that follow human genetic technology? The development of Islamic guidelines, rules and regulations was needed as a framework to tackle more and more possibilities that were yet to come.

Recent contemplation of bioethical matters in Islam has incorporated the utilization of the concept of *maqasid al-Shariah*. The *maqasid* seeks to protect five basic human interests – life, religion, property, intellect and lineage. This study will look into the details regarding the inquisition of human embryo research in the light of the *maqasid.*

The *Fiqh* and the *Maqasid al-Shariah*

What is the *maqasid al-Shariah* and how does it attend to the Muslim conception of the good and the ethical? The methodology of the Fiqh or Islamic law in essence provides the framework for moral deliberations on matters of social, ethical and legal connotations.

The Fiqh and the Shariah collectively signify Islamic law. Shariah means 'the path to the watering place' and can be defined as the clear path to follow in order to obtain 'guidance in this world and deliverance in the next '[2].

The first and primary source of the Shariah is the Quran [3]. Scholars believe that it may be referred to as the guidance of all times. The Sunnah, the ultimate code of conduct of the Prophet remains the next primary source of guidance for Muslims. The Hadith is a larger collection consisting of the early Sunnah and later compilations of the Prophet's sayings or acts observed and reported by the Companions and generations after the Companions [4, 5].

The importance of the Quran and Sunnah as the ultimate moral guidance is attributed to the Quranic verse, 'If you hold fast to them, never will you go astray after [my death]. They are: God's Book and His Prophet's Sunnah' [6].

Secondary to the Quran and the Sunnah are the *Ijma* and *Qiyas. Ijma'* denotes the consensus of opinions of the Companions during the period following the demise of the Prophet. Q*iyas,* or analogical reasoning from the Quran and the Sunnah, involves defining laws from a known injunction to a new injunction. A ruling may be compared to an established ruling from the Quran and Sunnah assuming that the new problem shares the same cause (illah) or essential reason for which it is applied [7]. Deductions regarding the problem can then be formulated and fatwas delivered. The effecting of the problem can then be graded as forbidden (haram) or permissible (halal) or obligatory or encouraged (harus) or not recommended (makhruh).

The method of *ijtihad* or independent logical reasoning is finally solicited to address matters not found in the Quran, Sunnah, Ijma and Qiyas. The process of *ijtihad* needs to consider the general purpose of the law or *maqasid al-Shariah*. Abu Ishaq Al-Shatibi (died 1388) cautions that those who neglect to make references to the *maqasid* become liable to errors in their *ijtihad* [8]. Three norms or subsidiary rules by which moral matters may be assessed were appropriately developed and these provide a practical scheme of Islamic moral philosophy:

1 The norm of compelling necessity (dharuriyyat)
2 The norm of needs or convenience (hajiyyat)
3 The norm of enhancement, Improvements or refinements (tahsiniyat)

Under the norm of *dharuriyyat*, the *maqasid* must be protected and in times of conflict the prohibition on an old ruling may be lifted:

'... He has explained to you what He has made *haram* [prohibited] for you except that to which you are compelled ...' (6:119) [3]

'[and] ... but if one is compelled by necessity, neither craving (it) nor transgressing, there is no sin on him; indeed God is most Forgiving, Merciful' (2:172–173).

Under the norm of *hajiyyat*, lawmakers are asked to exercise flexibility in deriving interpretations from scriptural sources for the purpose of eliminating hardships and the realization of benefits. Finally, the third norm *tahsiniyat* encourages acts of improvements and humans may seek beauty and comfort in life.

Acts of public interest implicitly include both matters of current and future worldly interests as well as the hereafter [9]. This would extend to actions taken to eliminate existing harms as well as anticipated harms. Permitting an action on the basis of unrestricted public interest is referred to as *maslahah mursalah* [3]. And it is within the realm of *maslahah mursalah* that human interests in particular, life, religion, intellect, property and lineage, are to be exclusively defended.

New Frontiers in Human Genetics and the *Maqasid al-Shariah*

Research on human embryonic stem cells produced from left over spare embryos through in vitro fertilization (IVF) procedures often raises concerns regarding the dignity of tiny 1-week old potential 'persons'. Many argue that if they have the dignity or moral status of persons they cannot be used as means but have to be treated as an end in themselves, even if the end is to relieve suffering of other persons. Writers have observed how arguments from human dignity are not featured predominantly in Islamic discourses on the ethics of human reproductive cloning [10]. In Western ethics, cloning is believed to undermine human dignity; this principle has been developed from secular humanism and Kant's categorical imperative and now entrenched within the 1997 UNESCO Declaration on Human Genome and Human Rights. According to Kant, humans are never to be treated as means but merely as ends in themselves. Genetic identity or uniqueness has been linked to the expression of human dignity and individuality. Cloning, which allegedly produces genetic copies of a person, would undermine that dignity.

Muslim opposition to human cloning is mostly shaped by concern over anticipated biological and social consequences. Most scholars lamented that cloning may cause disorder and interference in God's creation while some expressed fears of the idea that cloning produces copies of human individuals. However, they did not conclusively argue how this is unacceptable in Islam except that clones may create confusion in family lineage. Others echoed this concern by adding that the absence of the male component in procreation would not only produce disarray among family institutions, it would also contribute to the dilution of variations and cultural diversity among peoples. This is thought to be an antithesis to the purpose of grand creation: 'God has created you, males and females, of different creeds and races so that you may know each other'.

The importance of the principle of the protection of lineage was reflected in the quick issuance of fatwas regarding the status or legality of IVF births. The principle is derived from the primary scriptures: 'Among His signs is that He has created mates for you from among yourselves, so that you may find tranquility with them, and He has put love and compassion in you. Verily are signs for people who reflect' (Quran 30:11) [11], and, 'Marriage is my tradition. He who rejects my tradition is not of me' (Hadith).

Within the realm of the principle of the protection of lineage, surrogacy is quickly ruled out and frozen gametes and embryos are ordered to be destroyed in the event of death of one or both spouses, or in the event of divorce. But, the question of the morality of handling live entities (embryos) was not sufficiently examined. There was little inclination to bring about the attention that the personhood matter is being challenged here. Writers have noted elsewhere that scientific facets of the biotechnology are relatively least examined in discourses on IVF and embryo research among Muslim-led conferences [12]. This may be due to a lack of detailed understanding of how IVF tweaks the various processes within human embryology. This has produced rather poorly sustained deliberations of the moral status of human embryonic life and hence affects precise execution of the *ijtihad.*

Is early human life under threat here? To begin to answer this question, one needs to consider the importance of life in Islam. The sacredness of life and duty to protect life is evident in the Quran:

> 'Whoever kills a soul for other than manslaughter or corruption in the land, it shall be as if he killed all of mankind, and whosoever saves the life of one, it shall be as if he saved the life of mankind' (5:35).

We now need to examine if there is a definite time when the human individual is recognized to exist. There is, thus far, an assorted understanding of the beginning of human life across the lands of Islam. These ranged from direct scrutiny of Quranic verses singly or together with the hadiths to independent logical reasoning.

Muslim scholars often reflect how the first verse that was revealed to the Prophet (pbuh) was incidentally about the science of human embryology; 'Read in the name of God that creates. He has created you from a clot of blood' (96:1–2). Several places in the Quran further describe human creation and ensoulment:

> '... Then we placed him as a small quantity of liquid (nutfa) in a safe lodging firmly established. Then we have fashioned the *nutfa* into something which hangs (alaqa). Then We made *alaqa* into a chewed lump of flesh (mudgha). And We made the *mudgha* into bones, and clothed the bones with flesh. And then We brought it forth as another creation. So blessed be God, the best to create' (23:12–14);
>
> 'And breathe into him of His spirit' (32:9).

Most Muslim scholars agree that ensoulment occurs about 120 days after conception [13] while some scholars hold that it is much earlier at around 40 days [14]. The

Islamic Oath of Medical Ethics, however, offers a different perspective. It opens with a preamble, 'physicians ought to protect human life in all stages and under all circumstances, doing utmost to rescue it from death, malady, pain and anxiety' and, '... human life starts at the time of conception, and therefore cannot be terminated' [15].

To this effect, the Oath differs somewhat from other views on the beginning of human life in Islam. It seems probable that Al-Ghazali's view of embryonic development had influenced it. Al-Ghazali argued that termination of pregnancy at any stage designates a crime against the rights of the fetus. The severity of the crime increases as the fetus matures. He asserts that, 'the first is the part when the sperm lodges into the uterus. It next merges with the fluid of the women. It is at this point in time that it begins to receive life'. Islamic law has preset the penalty or blood money or *diya* for offenders and the amount simultaneously increases according to the age at which the fetus is terminated [16].

Past scholars advance the idea that the onset of ensoulment cannot be necessarily equated to the actual beginning of human life. For example, Ibn Qayyim (14th century AD) proposed that there is life before and after ensoulment. The early embryo, he said, has 'the life of growth like a plant but after ensoulment it gains the sense of perception and has preferences' [18].

The Quran and the Hadith have endlessly been studied in attempts to draw newer and perhaps better sustained analysis on the beginning of ensoulment. By juxtaposing scriptural knowledge against modern embryology, Aksoy [17] finds that 'the soul meets with body at the eighth week after gestation, around day 50 and forms the full human person'. Termination of the life of the embryo is considered *makhruh* or not encouraged before the 50th day and deemed as forbidden after this stage [17].

The derivation of fatwas on the beginning of human life often illustrates how *ijtihad* is performed relative to local cultural needs. For example, when Indonesian Islamic scholars in the year 2000 announced that 'life begins at 120 days after fertilization', Malaysian officials opted to take the stance that 'life begins at the first moment of conception'. Research findings by this writer showed that the latter decision was made in view of prevailing social problems – to curb the incidence of illegal abortions. Circumstantial evidence, which affects public interest, allows a ruling to deviate from the general consensus that life begins at 120 days [19].

The dynamism of the Shariah is also illustrated when the Mecca Council of Islamic Jurisprudence issued a fatwa that overturned older fatwas on abortion. The Council arbitrated that abortion is allowed if the fetus is found to have gross malformations and is less than 120 days old [18]. Earlier fatwas embark from the premise that abortion is allowed if and only if the mother's life is at risk. Subsequently, moves to ensure non-defective births have induced scholars to highly recommend genetic counseling and prenatal genetic diagnosis for the fact that decisions for termination of pregnancy can be made at earlier stages or way before pregnancy is established. It may therefore be argued here that similarly stem cell research to detect and eliminate disease-laden genes may appropriately be permitted.

The plurality of views on the beginning of life within Islamic circles would make it difficult to bring a closure to the ethics of stem cell research. There are scholars who steer clear of hinging their debates on the actual moment when life exists to the premise that potential life is imminent once fertilization occurs. The legal verdict used to acknowledge activities related to IVF or cloning or stem cell research is not strictly 'not permissible or not *halal*' but the more subtle ruling, 'not encouraged or *makhruh*'. This demonstrates cautious reflection – that there is potential life and tampering with it cannot be declared as outrightly permissible.

Recent scholars offer novel though ambiguous ideas which suggest that the embryo may have the right to inheritance but not necessarily the right to life. It is held that,

> 'the human embryo and fetus must be treated with dignity in conditions appropriate to human dignity and therefore the number of embryos may not be higher than that which ensures a good chance of successful treatment. [But] embryos should not be treated like 'property' and cannot be subjected to marketing and financial gain ... it has the right to legal inheritance [but] it cannot be clearly stated that the embryo has a 'right to life' the same way the unborn child has the right to life' [20].

Other scholars hold rather radical positions on the status of the embryo grown in vitro. They note the Quran describes the creation of a human being as forming within 'three folds of darkness' namely the uterus. The absence of the uterus in the case of IVF embryos supposedly takes away the moral rights of that embryo. This idea forces the significance of the womb as an essential nurturing environment for the formation of a 'complete' human being. The embryo in the laboratory cannot morally be equated to the fetus in the maternal womb. The former has no rights and therefore no moral status and deserves no respect or protection. This proposition was raised during a meeting of the Islamic Organization of the Medical Sciences [21]. To this day, it is unclear if this forms an authoritative stance that may subsequently be extended to support the creation and utilization of stem cells. This would dismiss protests over the use of in vitro embryos as objects.

The derivation of rules that permit organ donation by Muslims is believed to be motivated by the principle of 'prevention of hardships and the realization of benefits'. 'God intends you comfort and ease and He does not want to burden you with difficulties' (1:185). In addition, the philosophy of illness and medicine in Islam is supported by the *hadith* which states that 'No disease is created whereby He has not also created the cure. There is cure for all diseases except death.' Thus in Islam concepts regarding sickness and medical treatment and medical research is constructed within a faith framework. It subscribes to the act of submission to an all-powerful God and the dependency of the doctor and the scientist on God's order. This may also be taken to be the general Islamic stand when deliberating on the expansion of medical endeavors into the line of stem cell technology.

This philosophy has been seen in Islamic bioethics which dismisses fears that human cloning constitutes an attempt to 'play God'. In 1997 the Islamic Fiqh Council declared that cloning cannot be interpreted as an act of playing God. In the same way that a person sows the seed of the plant and God wills it to form into a plant, the same way must we interpret the act of cloning. After all sowing a seed is only the 'cause' and God will produce the 'effect' [17]. In the same line of reasoning, manipulating stem cells to enhance human genetic traits with the view to eliminate diseases cannot be likened to an intervention in God's creation and God's plan.

Are humans absolute owners of their body parts including cells and tissues? Must consent be retrieved from 'owners' before they can be used or researched on? The concept of *tauhid uluhiyyah* deals with the belief in the unity of God and everything that is in the universe are His alone. In the context of ownership, humans are mere trustees to all resources available to them. As trustees, they are responsible and must see to it that such resources are shared equitably within their community. In discourses on the ethics of organ donation in Islam, donors would surrender ownership of organs upon their death. These organs are to be managed by the medical institutions to be utilized as optimally as possible, 'And spend of your substance in the cause of God, and make not your own hands contribute to (your) destruction; But do good for God loveth those who do good' (2:195).

The same argument may be advanced to the problem of ownership of stem cells. Since the human individual has no rights to his/her cells, it may be used in research. However, regulations must be established to ensure that spare or research stem cells are not used for reproductive purposes such that it transgresses the *maqasid,* family lineage. This is clearly found when Muslim countries urge that stored gametes and embryos be destroyed if one or both owners of such entities were to die or if the couple divorces.

The rule of moderation applies to a Muslim and has even been extended to the *ummah* or society as a whole – striving to be a moderate society, *ummah wasatah,* is a moral goal in Islam [22]. Excessiveness in lifestyles and asceticism in religious obligations are proscribed in Islam. Does this mean that striving for perfection, enhancement and improvement in livelihood is prohibited? For this Hathout offers some thoughts, 'stem cell research on pre-implantation embryos may be justified if the aim is to save actual persons suffering serious illness, on the basis of the choice of the lesser of two evils' [23]. Stem cell research is recognized as an evil, the lesser one, and may only be used sparingly in medicine but not for overzealous intentions to create perfection in humans.

The *maqasid,* the protection of the intellect, compels Muslims to seek knowledge and enhance their understanding of the world. 'Behold! In the creation of the heavens and the earth, and the alternation of night and day, there are indeed signs for men of understanding' (3:190) [11]. Knowledge is not only sought of Muslims to increase their understanding of the world they lived in but also believed to sustain their faith in the religion. The Islamic Oath of Medicine guarantees that 'the pursuit of knowledge has double value in Islam – it is both therapeutic … and … an act of worship' and 'God will raise the ranks of those of you who believed and those who have been given knowledge'.

The need to grab back the glorious days of Islam in the sciences when great names such Avicenna, Averroes, Ibnu al-Haitham (Al Hazen) and Al Biruni reigned, is often chanted to motivate researchers [24]. Scientists are persistently asked to reenact the golden era of the Islamic sciences, a symbol of the intellectual value of knowledge. Endeavors in genetic research therefore conform to the *maqasid* and it has earlier been argued that manipulation of stem cells in research does not transgress the notion of 'property' because humans do not own them nor does it imply immoral manipulation of 'life'.

Conclusion

Moral assessment of genetic technologies may be made within the exclusive light of the *maqasid al shariah.* For the good of the public, research to bring about enhancement of the quality of human life may properly be encouraged and possibly assist in the protection of life, lineage, intellect, property and religion. The ethics of genetic advancements and their potentials must primarily rest on the principle of upholding the family lineage and the sanctity of the marriage institution. Discussions on the beginning of human life have been diverse but there is general consensus that there is potential life in early embryos and they must be treated with caution. The intention to eliminate diseases is justified in actions that may bring about the possibility of embryo destruction. This is interpreted to be the lesser of two evils. By recognizing that the act of destroying embryos is evil is sufficient to show that there is recognition among scholars of Islam that embryos do possess moral status no matter what age.

Ibn Qayyim once uttered these words, 'the interpretation of the law should change according to changes in time, space, circumstances and tradition [and] neglecting these will bring injustice to the Shariah and entails more harm and difficulties'. In view of protecting the interest of the human being, such consequences would never be tolerated [25]. A reasoned and sustained deliberation on the ethics of stem cell research is further needed. Failure to execute rigorous *ijtihad* and sufficient consideration of the needs of the people will make religion insignificant in this biotechnological era. It will be most unfortunate for religions if biotechnologists continue to claim;

> 'Scientific evidence concerning fertilization and the development of the very early embryo [will] make a purely philosophical or theological stand untenable' [26].

References

1 Dabu C: Stem cell research stirs debate in Muslim world, too. Christian Science Monitor, June 22, 2005.

2 Kamali MH: Nature, sources and objectives of the Shariah. Islamic Q 1989;33:215.

3 Kamali MH: Principles of Islamic Jurisprudence. Petaling Jaya, Pelanduk, 1989, p 17.

4 Rahman F: Sunnah and hadith. Islamic Studies 1962;1:4.

5 Iqbal M: The Reconstruction of Religious Thought in Islam. Lahore, Iqbal Academy and Institute of Islamic Culture, 1989, p 171.

6 Ramadan S: Islamic Law, Its Scope and Equity. Kuala Lumpur, Muslim Youth Movement of Malaysia, 1987, p 24.

7 Fyzee AAA: Sources of law; in Cumberlege G (ed): Outlines of Muhammadan Law. London, Oxford University Press, 1955, p 16.

8 Al-Qaradawi Y: The lawful and the prohibited in Islam. El-Helbawy K, Siddiqui MM, Shukry S (transl.). Kuala Lumpur, Islamic Book Trust, 2001, p 49.
9 Kamali MH: Shariah and the challenge of modernity. J Malay Inst Islamic Understand 1994;2:1–26.
10 Eich T: The debate about human cloning among Muslim religious scholars since 1997; in Roetz H (ed): Cross Cultural Issues in Bioethics: The Example of Human Cloning. Amsterdam, Rodopi, 2006, pp 311–340.
11 Yusuf Ali A: The Holy Quran: Translation and Commentary. Jeddah, Dar Al-Qiblah, 1982.
12 Sachedina AA: Culture and religion in Islamic biomedicine. The case of human cloning; in Roetz H (ed): Cross Cultural Issues in Bioethics: The Example of Human Cloning. Amsterdam, Rodopi, 2006, pp 291–310.
13 Albar MA: When is the soul inspired?; in Contemporary Topics in Islamic Medicine. Jeddah, Saudi Arabia Publishing & Distributing House, 1995, pp 131–136.
14 Kamel AA-A: Abortion; in Al-Gindi AR (ed): Seminar on Reproduction in Islam. Kuwait, Islamic Organization for Medical Sciences, 1983.
15 Athar S: Islamic Perspectives in Medicine: A Survey of Islamic Medicine: Achievements and Contemporary Issues. Indianapolis, American Trust, 1993, p 195.
16 Al-Ghazali AHM: Marriage and Sexuality in Islam. Farah M (transl). Salt Lake City, University of Utah Press, 1984, p 110.
17 Aksoy S: Making regulations and drawing up legislations in Islamic countries under conditions of uncertainty, with special reference to embryonic stem cell research. J Med Ethics 2005;31:399–403.
18 Albar MA: Ethical considerations in the prevention and management of genetic disorders with special emphasis on religious considerations; in Majeed AB (ed): Bioethics in the Biotechnology Century. Kuala Lumpur, Institute of Islamic Understanding Malaysia, 2002, pp 107–122.
19 Nor SNM: Ethical issues in artificial reproductive techniques; in Mohd Salleh SMSS (ed): Genealogy and the Preservation of the Progeny. Kuala Lumpur, Institute of Islamic Understanding Malaysia, 2006, pp 107–146.
20 Elcioglu O, Yildirim A: Ethical and legal problems with assisted reproduction in Turkey. J Int Soc Hist Islamic Med 2004;3:33–38.
21 Al-Khateeb SE: General comments on sexual assault, fate of bank-deposited embryos and surplus fertilized ova. 3rd Symp Medical Jurisprudence – Islamic Vision of Some Medical Practice. Jeddah, April, 1987.
22 Al-Faruqi I: On the metaphysics of ethics in Islam. Listening: J Culture and Religion 1979;14:25–43.
23 Hathout H: International Seminar on Human Genetics and Reproductive Technologies: Comparing Religious and Secular Perspectives. Cairo, Islamic Organization of Medical Sciences, 2006.
24 Zin SM: Towards a quantum leap in the development of Islamic science in Malaysia; in Abdul Razak A, Abdul Majeed AB (eds): Islam, Science and Technology. Kuala Lumpur, Institute of Islamic Understanding Malaysia, 1997, pp 47–67.
25 Mahmassani S: The Philosophy of Islamic Jurisprudence in Islam. Farhad Ziadeh (transl). Kuala Lumpur, Hizbi Publications, 1987, p 107.
26 Abdullah MF: Reproductive human cloning: medical, scientific and strategic perspectives. Seminar on Reproductive Human Cloning. Kuala Lumpur, Ministry of Foreign Affairs and Institute of Strategic and International Studies Malaysia, February 2002.

Siti Nurani Mohamed Nor
Department of Science and Technology Studies
University of Malaya
Kuala Lumpur (Malaysia)
Tel. +60 3 79674166, Fax +60 3 79674396, E-Mail sitinuraninor@um.edu.my

Pfleiderer G, Brahier G, Lindpaintner K (eds): GenEthics and Religion. Basel, Karger, 2010, pp 138–150

Buddhism and Human Genome Research

Pinit Ratanakul

College of Religious Studies, Mahidol University, Nakornpathom, Thailand

Scientific revolution in genetics began with the discovery of genes and DNA, the building block of life, followed by the use of genetic engineering technologies to manipulate genes of the organism, and culminating in the establishment of the Human Genome Project to map genes and sequence human DNA. This scientific breakthrough is a boon for humanity. It opens up possibilities for human beings to be healthier and live longer. Advances in human genetic research and its applications in medical practices such as diagnosis, treatment, and prevention of genetic diseases are of great promise and bring hopes for the cure of incurable diseases which many people are afflicted with. However, the new medical treatment and technologies raise many crucial ethical, legal and social issues that society's existing morality and legal system seem unable to deal with adequately. The ongoing international discussions of these issues express the need to clarify the moral implications and human values at stake in scientific research and medicine, as well as to find solutions that reflect high human values.

These emerging ethical issues are also new challenges to religion which was not prepared to cope with them. In order to meet these challenges religion has to be engaged in ethical reflection on the issues and to find solutions in accordance with its teaching, and relevant to the situation. Only recently has Buddhism been involved in such a reflection to contribute its insights, value perspectives, and concerns to international discussions of these issues. The purpose of this chapter is to present Buddhist ethical reflections on human genetic research and its implications for the future of humanity. Before dealing with these reflections, it is to be noted that Buddhism is an ally of science. It has an open attitude to scientific research and also has an ethic that is applicable to it. With such preliminary understanding one can appreciate the Buddhist position regarding ethical issues raised by scientific research. Accordingly this chapter is divided into three parts. The first part is a general observation of the relation between Buddhism and science, while the second is a short account of Buddhist ethics, followed by the section about reflections on human genome research and on the future of humanity in relation to the progress of science and technology.

Buddhism and Science

There are close analogies between Buddhism and science. An obvious analogy is the Buddhist view of the universe as uncreated, an orderly self-sustaining process regulated by the natural law of cause and effect *(Dhamma-Niyāma)*[1], also known as the law of conditionality *(Paticca-samuppada)*, stating the relationship between causes and effects mutually conditioning and conditioned. This self-existing law is a cosmic law *(niyāma)* manifesting as the physical law *(Utu-Niyāma)* and the biological law *(Bija-Niyāma)* in the material realm, as the psychological law *(Citta-Niyāma)* and the moral law *(Kamma-Niyāma)* in mental and moral domains. The causal law and all its manifestations can be discovered by man through his reason and intuition.

Buddhism is also analogous to science in the search for truth to alleviate human suffering. In such search Buddhism and science each devises its own method of investigation in accordance with the kind of truth it seeks. Science strives for the knowledge of physical reality in an objective way, while Buddhism uses the first person method to investigate the inner world of the mind to discover the causes of suffering and the release from it.[2] Similar to scientific investigation, the Buddhist introspection explores the mind in its reality without any preconceived idea or sentiment. In this sense it is more an objective analysis of the mind, and its truth so gained is verifiable. Both Buddhism and science see the need for the verification of truth by reason and experience. Reason is necessary especially to test the truth that goes beyond mere facts, whereas experience gives certainty to the truth so arrived at. The Buddha always claimed that the dhamma, his teaching, is not unsupported religious dogma to be believed in faith, but is open to criticism and objective analysis. He usually advised his disciples to accept the dhamma only after verifying that it is in accord with reason and reality. The Buddhist systematic meditation is offered as a means of such verification by enabling the mediator to see truth 'face to face' *(ehipassiko)*. Perhaps, it is this free and open spirit in Buddhism and its verifiable truth that have attracted many intellectuals and scientists to this religion. Einstein himself wrote that if there is any religion which is acceptable to the modern scientific mind, it is Buddhism. This open mind underlines Buddhist responses to scientific research. The responses are more from reasoned analysis rather than from the narrow angle of dogmatic teaching.

Buddhist Ethics

The sole concern of Buddhism is suffering and its cessation.[3] It was with such concern that Prince Siddatta of Sakka's Kingdom in northern India left his palace and the self-

[1] Pali terms are used throughout the paper.
[2] For more discussion on the introspective method, see Wallace [1].
[3] 'This I do teach', declared the Buddha 'suffering and cessation of suffering' [2].

indulgent life to be a wandering ascetic and, after 6 years of intense striving, became the Buddha, the Enlightened One, and found the way for absolute emancipation. In *Dhammacakkappa-vatanasutta*, the first sermon that established the religion of Buddhism, the Buddha expounded to the five ascetic disciples the Four Noble Truths *(Ariya-sacca)*, discovered in his Enlightenment *(bodhi)*, i.e. the reality of suffering *(Dukkha)*, its root-cause *(Samutthāya)*, the cessation of suffering *(Nirodha)*, and the Path leading out of it *(Magga)*. This Path is referred to as the Eightfold Noble Path *(Ariya-Magga)* with its threefold divisions: morality (*sīla*), concentration *(samādhi)*, and wisdom *(paññ)*.

The goal of Buddhist ethics is to help the individual become a good person leading a virtuous life, the morally worthwhile existences. With this purpose Buddhist ethics emphasizes and analyzes the traits of character or virtues *(dhamma)* that individuals ought to develop and possess in themselves to become virtuous persons. At the same time it formulates normative principles or moral rules to be adopted as guides to moral actions. This special concern with virtues and moral rules is the basis of Buddhist ethics. It gives equal attention to ethical principles for right action, and the person behind the action, his qualities, motivations and intentions. Cultivating virtues in one's heart is an important way of becoming virtuous. However, we cannot read the heart entirely, but actions can be observed and judged. We usually assume that good people perceive what is morally right to do and do it, but this is not always the case. From experience we know that very often good people fail to perceive what is right to do in a given situation and face a conflict in which they do not know what ought to be done. The adoption of moral rules will enable them to make the right decision. Following moral rules is then another important way of becoming virtuous.

The core of Buddhist ethics is compassion *(karunā)*. It is the basis of all good actions. It involves beneficence, non-maleficence, and other forms of altruism such as benevolence and generosity. Compassion in Buddhism is the quality of the heart and is not an ephemeral sentiment. It is a sublime emotion[4] arising from the knowledge of suffering inherent in all sentient beings, and also from the realization of their unsubstantial nature *(anattā)* and inter-relatedness *(paticcasamuppada)*.[5] Breaking through the veils of ignorance *(avijjā)* and delusion *(moha)* of the enduring self we are brought to the bond of fellowship with others.[6] This realization of

[4] There are 4 sublime emotions *(Brahma-Vihāra)* i.e. loving-kindness *(mettā)*, compassion *(karunā)*, sympathetic joy *(muditā)*, and equanimity *(upekkhā)*.

[5] *Anattā* also means 'not-self' or 'non-ego' as oppose to *attā* or self-existing ego-entity. It is one of the Three Characteristic of Existence *(Ti-lakkhana)* taught by the Buddha, i.e. suffering *(dukkha)*, impermanence *(anīccā)* and non-substantiality *(anattā)*. *Paticcasamuppada* refers to the Buddhist doctrine of Dependent-Origination stating the conditionality and dependent nature of all physical and psychological phenomena.

[6] *Avijjā* or ignorance means the lack of knowledge about the true nature of all phenomena, i.e. *dukkha, aniccā*, and *anattā*. It is also referred to as not knowing the Four Noble Truths, i.e. suffering, its origin,

oneness with others is the springboard of love and compassion. Since compassion arises out of knowledge, it implies justices as the mind, free from the illusory separate identity and self-seeking impulse, is fully open to the multifarious needs and perspectives of others. With such an open mind we can see the dignity of other people, and are able to accept them as they really are and respect them in their true being.

Compassion therefore enables us to treat others as equals, as full persons in their wholeness, and as fellow members in our common humanity. Hence compassion has a social character as it recognizes the moral equality of others, giving to and claiming for them what is due to one another as fellow human beings. But compassion can and does go beyond justice and therefore beyond equality and human rights. It is voluntary sacrifice of one's claims or rights beyond what is socially obligated. Compassion then has special application in the relations between human beings, and between human beings and nature. It is both the moral value that good people possess and a moral rule for right actions.

Equally emphasized in Buddhist ethics are human action and human freedom. These are explained in relation to the moral law, which is referred to as law of *kamma (Kamma-Niyāma)*, the life-cycle *(samsāra)*, and absolute emancipation *(nibbāna)*. As already mentioned, the moral law is a part of the cosmic law of cause and effect *(Dhamma-Niyāma)* operating in all domains ranging from nature to morality. This moral law affirms the correlation between actions *(kamma)* and consequences *(vipāka)*, and acts as a cosmic force that determines the appropriate sanction of one's action.[7] Good actions produce good consequences, and vice versa. Everyone has to reap the result of his own sowing. Even the undetected criminal cannot escape the effects of his bad deeds as life is not limited to a single individual existence. Present life is only a part of an endless life-process of arising (birth), passing away (death), and again-arising (rebirth), stretching out across space and time, specially referred to as *samsāra*, or the life-cycle of death and rebirth. In this *samsāra* existence there is a hierarchy of living beings ranging from gods, humans, animals, hungry spirits to hell-beings dwelling in different realms. Despite differences in lifespan, all of them are subject to suffering and rebirth and migrate from one realm to another according to the results of their own deeds *(kamma)*, good or bad.[8]

Of all rebirths human rebirth is the most precious and the most difficult one. This is because only the human has the potentiality to free himself from *samsāra* existence

and the way to its cessation. *Moha* or delusion is one of the three unwholesome roots of suffering. It is a synonym of *avijjā*.

[7] *Kamma* means intentional action, bodily, verbal, and mental, which has consequences on the doer.

[8] In Buddhist cosmology there are 3 realms of rebirths (*loka*) in *samsāra*, i.e. the realm of desire (*kammaloka*), the realm of forms *(rūpaloka)*, and the formless realm *(arūpaloka)*. The first realm consists of the higher spheres of gods the middle spheres of sentient beings, humans, and animals; and the lower sphere of ghosts, spirits, and hell-beings. The realm of forms and the formless realm are the abodes of the most refined and subtle being *(brahman)*. For more details, see Sadakata [3].

and thereby to attain absolute emancipation.[9] Other suffering beings in the lower spheres will gain precious human rebirth only when the effects of their bad *kamma* that led them to their lower rebirths are exhausted. When this happens the results of the good deeds done when they were human will lead them to human rebirth right away or to rebirth in other forms of life, and, sooner or later, to human rebirth again. The rarity of being born human is illustrated in a parable given by the Buddha. It is about a partly blind turtle in the vast ocean where there was a piece of floating wood with a hole in it. The Buddha estimated that the chance of any particular being to obtain a human rebirth is as about equal to the likelihood of that turtle pushing its head through the hole in that piece of floating wood. Due to the preciousness of human rebirth the first precept or moral law in Buddhism prohibits the taking of life, particularly human life, in any stage of development. The precept is compassion in action involving non-maleficence or doing no harm to other people. It also has social and religious significance. The religious consequence is crucial with regard to rebirth. Those who keep the precept are assured of good rebirth. Its transgression yields bad kammic consequences such as brevity of life, ill health, and rebirth as hell-beings with irremediable suffering.

Buddhist ethics is humanistic. It centers on man and his deeds which have far-reaching effects on the individual in *samsāra* existence. Good deeds lead to higher rebirth as gods or humans whereas evil deeds cause lower rebirth as animals, hungry spirits or hell-beings. This is because man is not created, and his life is not predetermined. He is the total sum of his actions. The imprisonment in *samsāra* existence is neither arbitrary nor mandated by a tremendous power. It is rather the results of one's own deeds, good or bad. Through his deeds each man weaves his own web of fate. He is both cause and effect, the entire act or deed on the one hand, and the consequence of the act on the other hand. As an *effect* he is conditioned by past deeds. What he is now is the result of what he has been before. But as a *cause* he is a field of possibilities. The present is the interface of past and future. It is quite brief and fleeting. But it is in this elusive present that he is the chooser and the doer as he has the freedom and ability to gradually liberate himself from the past to become whatever he wants to be in this life and after life, and above all, to sever the knot that ties him with *samsāra* existence.

This freedom is both the glory and the tragedy of man. It can take the individual out of suffering or increase suffering in the *samsāra* existence. Virtues and moral rules are necessary for the development of moral character and for the making of right actions that have good kammic consequences on his present and successive lives. Buddhist ethics is specifically formulated to help the individual to make good use

[9] It is because the human sphere is the only place where there is enough suffering to motivate humans to seek ways to transcend suffering and enough freedom to act on their aspirations. In the higher sphere the gods are too absorbed in the blissful state to find ways out of samsāra existence while animals, spirits and hell-beings are in irremediable suffering and have little freedom to do good or bad.

of his freedom by replacing the negativities of ignorance *(moha)*, greed *(lobha)*, and anger *(dosa)* with knowledge and compassion. The negativities are the root-causes of suffering. When knowledge and compassion form the bases of thoughts and actions one will live a truly human life with happiness.[10]

The Human Genome Project and Human Values

Buddhism sees the Human Genome Project and genetic engineering technologies as important ways to relieve human suffering from disabilities and diseases as well as to promote human wellbeing. Though these goals are beneficial to humanity they cannot be used to justify the procedures of the research when they are unethical. Human genome analysis needs to be made in harmony with essential human values of human rights and compassion. With this harmony, the human subjects will be protected and the research, or science in general, will be conducted to meet its humanitarian ends.

In Buddhist ethics, moral issues and moral obligations are analyzed in terms of duties and responsibilities. The concept of human rights, originated in the liberal democratic reformist philosophies in the West, has now become a universal term of reference in ethical discussion and in talking about our common humanity and moral equality. There is no equivalent expression in Buddhism.[11] Until recently the full meaning of the concept was absent from Buddhist ethical thinking and moral norms. Within Buddhist ethical framework the fundamental human rights needed to be strictly respected in human genetic research are the right to life of the embryo and its right to self-determination. The question of the personhood of the embryo, usually raised in Western bioethics, does not arise in Buddhist ethics because, in its view, the embryo is human and thus possesses human dignity and human rights at the time of conception, and not after the 14th day as upheld by the 'person' theory.[12] This Buddhist view is also shared by the 'soul' theory though on different grounds. In this theory the humanity of the embryo is the result of its ensoulment by God at the very moment of fertilization, whereas in Buddhist understanding, human life begins with the infusion of the *gandhabba*, the individual kammic life-force into the womb.[13] Even when human life manifests itself in the minute form (*kalala*) as a fertilized egg,

[10] For further views of Buddhist ethics, see Keown [4].

[11] In Buddhist ethics compassion is a comprehensive value embracing all other crucial human values such as justice, human rights, and human dignity.

[12] This widely held theory asserts that before the 14th day of its development the embryo is not a person or a human being and thus does not possess human rights. The theory is used to justify the destruction of the embryo in its earliest stage to obtain stem cells for research.

[13] Here *gappcuti* refers to a suitable individual life-force ready to manifest itself in a particular womb. In Buddhism the concept is used only with this meaning, and must not be mistaken for a permanent soul which is diametrically opposite the Buddhist *anattā* doctrine.

it is still precious and is entitled to be respected and protected. Therefore it is unacceptable to use and destroy the embryo as experimental material.

Equally important in Buddhist ethics is the embryo's right to self-determination. This right is congruent with the Buddhist teaching of *kamma* which puts total responsibility on the individual for his life. He has the freedom and ability to make his own choice of values, and to plan his life according to his perspective of what is good and right. These freedoms and abilities are the human potentials of the embryo and form the base of its dignity. In research involving human subjects, they are understood in terms of informed consent, voluntarily given by the individual participating in the research. The respect of these two essential human rights will protect the embryo from exploitation in the name of scientific progress. However, in today's interdependent world where science and society are mutually dependent and where science has the power to shape society, human rights, particularly the right to information and the related value of informed consent, should not be narrowly defined in terms of individual rights alone. Their meaning and scope should also be extended to include the public or the community as necessary. This extension is particularly needed with regard to human genetic research. In this case the public as a whole, not merely isolated individuals, has the right to be informed about issues relevant to their own welfare. Similarly, informed consent should be enlarged from the setting of two individuals meeting face-to-face to become more an interaction between the scientists and the public at large. Consent can be secured through public debate before launching a project that has major impacts on society and its future. Giving full and complete information to the public about the potential damages of research without exaggerating possible benefits of the research, and without minimizing its potential harms, is a moral obligation and social responsibility of scientists.

Genetic Manipulation and Human Dignity

The new capacities of science to interfere with the natural selection and other natural processes through genetic manipulation have been considered by some ethicists as 'unnatural' and not 'normal', and therefore ethically unacceptable. This raises the question of what is to be considered 'normal' and what is 'naturally given'. The answers depend on our definition of nature and our understanding of its laws. In Buddhist perspectives nature is uncreated, and much more comprehensive than our understanding. Its secrets are mysteries to an unenlightened mind. As a part of nature, the human being is also uncreated, and like all living beings, has no fixed identity or essence from birth. Man is what he is not, and is not what he is. The change in genetic individuality is not against nature. Anything that is against it cannot happen. Everything is impermanent and has potential for change. Besides such change itself is not important and worthy of ethical consideration. What matters most are the effects of this change on the individual, society, and humanity. For being human, ethical integrity is more important.

Genetic manipulation is ethically acceptable only if it helps release the suffering of an individual and promotes human well-being in general. With such consideration, in Buddhist ethical thinking only gene somatic cell therapy for medical purpose is acceptable because of its potential benefits to cure genetic diseases and other ailments. However, since the technologies for this therapy are not yet perfect, every effort must be made to ensure the safety of the patient. Accordingly research proposals for somatic cell therapy need to be carefully assessed given special attention to life-threatening diseases that cannot be treated by conventional methods. On the contrary the use of gene somatic cell therapy in cosmetic medicine for self-enhancement is ethically unacceptable due to its potentially negative social effects. Only the rich can afford its high cost and use it for their own advantages over others who cannot afford it. This will aggravate the existing disparities between the 'haves' and the 'have-nots', and increase greed and self-preoccupation among them. Similarly germline therapy involving the reproductive cells which can be inherited by all succeeding generations is also ethically impermissible due to its possible negative effects on society and human species. The parents' desire for their offspring to have the most desirable genetic makeup is everyone's dream. But should science develop to meet this desire? The concern with the welfare of the yet-to-be-born is good intention. However, it can further lead to an eugenic movement to develop ideal human beings, free from all defective genes. This can be similar to what the Nazis did in their eugenic programs to produce perfect human beings. If this happens, genes will become the determining factor for discrimination. We will then be living in a society in which there are privileged people with good genes and those with undesirable genes who can be socially and legally obligated not to have any offspring. Those with suspected defective genes will then be prevented from being born. Should we allow our society to be like that?

Human Cloning and Stem Cell Research

Human cloning is also another scientific intervention in the process of reproduction to provide a new human being who is genetically identical to an already existing human. Like genetic manipulation human cloning has been looked upon as against nature, and considered as playing God. In Buddhist thinking, if the making of human life through this method is really in violation of nature it will not be successful, and the clone cannot come into being. It is also a matter playing God. This is because, according to the law of conditionality, everything arises and exists due to the presence of certain conditions, and ceases once these conditions disappear. A flower, for example, comes into being because there are sufficient conditions such as seed, soil, moisture, air, and sunshine. If one of these factors is absent there will be no flower. Similarly, life cannot be created. It manifests itself when there are sufficient conditions. In the cloning process the scientist acts only as an agent to arrange the sufficient conditions to bring the clone into being. It is *gandhabba,* the individual kammic life-force, which will decide if it will be born under such conditions.[14]

Buddhism does not support reproductive cloning though it is sympathetic to those who have lost their loved ones and want to have clones in their places. In its view attachment to the substitute is not the right way to cope with such suffering. Though the clone and the original are genetically identical they are different in mentality and character. The clone can either decrease or increase the suffering of the parent. Whatever the situation is, one cannot escape the suffering caused by death. Other forms of human suffering that are unavoidable are old age, sickness, contact with what one dislikes, separation from what one likes, and not getting what one wants.[15] All these make up the painful side of existence which humanity tends to ignore. The Buddhist way to deal with suffering from the loss of loved ones and suffering in general is not to find substitutes in pleasure-giving areas such as sex, drugs and liquor as they feed ego-gratification, the root-cause of suffering. Rather it is to accept the reality of suffering and to live with it wisely. There is no balance of happiness[16] and suffering in human existence. The painful side of experience usually outweighs happiness. However suffering also has a positive side. It makes us aware of the vulnerability of life. Health or human wellbeing does not necessarily mean the absence of suffering, but the ability to deal with it and to use it and to transcend it in the name of personal growth and sympathetic understanding of others.

Apart from this wisdom one should think about the potential effects that reproductive cloning has on men, women, family, society and humanity, as well as on the clones themselves. Though there may be some positive effects, they are outweighed by the negative ones. These include the change in meaning and place of parenthood, the abuse of cloning, the commercialization of cloning, and the health of the clones, physically and mentally. Such possible negative effects will add more suffering to what we already have. In the light of these negative possibilities generating cloned offspring is not ethically permissible.[17]

The Buddhist emphasis on the value of life or, in the language of rights, the right to life of the embryo and on the need to protect it does not facilitate embryonic stem

[14] It is stated 'when this exists, that exists, when this arises, that arises, when this is not, that is not, when this ceases, that ceases' [5].

[15] These are pointed out in the First Noble Truth.

[16] The preoccupation with suffering has made many non-Buddhists regard Buddhism as a pessimistic religion. Such misunderstanding arises from their looking at suffering apart from the context of the Four Noble Truths. The Buddhist approach to suffering emphasizes a sense of realism and certainly discourages any attempt to brood over suffering and be agonized by it. It would be more appropriate to describe Buddhism not as 'pessimistic' but as 'realistic' for it begins with the fact of experience. However, the Buddha's teaching went far beyond this simple observation of the facts of life. He penetrated the causes of suffering and showed the way to end it.

[17] For further discussion of human cloning from Buddhist perspectives, see Ratanakul [6].

(ES) cell research. Though the goal of the research is altruistic, e.g. to find a remedy for a particular disease, in the procedures the destruction of the embryo to extract stem cells is unavoidable. ES cells are all purpose cells with great potential for curing incurable disease and other ailments. The destruction of the embryos to obtain stem cells is a controversial issue. There are pro and con arguments. The religious prohibition on the use of ES cells in gene therapy to cure incurable diseases may be regarded as not being concerned for the cure of the ailment of those who are suffering. Despite its spirit of openness and the lack of strong dogmatism in its teaching, the Buddhist position on this issue is uncompromising. Killing is killing, no matter for what purpose. It is our moral obligation to protect the embryo and provide it with opportunities to realize its human potentials. One should not forget that never before in the history of humankind has human life been threatened by science as in our times. What is destroyed in the ES cell research is real human life with all its potentials, and not life in statistics or in books.

This Buddhist stance is challenged by the case of excess embryos from in vitro fertilization that are kept frozen in the fertility clinics, and which are to be discarded because they exceeded their storage life. These excess embryos are also human and their human rights are to be protected. Is it acceptable to use them to obtain ES cells for scientific research? In Buddhist ethics voluntarily giving of one's life to save the life of the other is a good kammic action. But since these surplus embryos are not in a position to decide and to give consent is it morally permissible if we decide for them to sacrifice their lives for the interest of scientific research? Is such a decision made in their best interest? The answer may be yes or no. Each has its rationale. In this kind of case in which the line between 'right' and 'wrong' is blurred, Buddhist ethics leaves it to the individual to decide according to his conscience.[18] If the Buddhist breaks the first precept, e.g. killing any excess embryo to obtain ES cells for experimental use, he does so with full awareness of the kammic consequences on him.

The gravity of the result of the result depends on many factors such as the intention behind the act, the intensity of the doer's intention and effort as well as the size and quality of the being that is killed. The development of the embryo implies different degrees of the potential of the embryo which itself influences the kammic consequences.[19]

The Buddhist precept against killing and all other percepts are ideals to live up to, and not 'commandments' given by the Buddha to be believed though they were taught by him. The Buddha was a teacher and not a lawgiver. The precepts are the

[18] Passive euthanasia is another case [7].

[19] In Buddhist text it is stated that for an act of killing to be complete and kammically potent (i.e. producing evil effects), five conditions must be fulfilled: (i) the presence of a living being, human or animal; (ii) the knowledge that being is a living being; (iii) the intention to kill; (iv) the act of killing by appropriate means, and (v) the resulting death. In the absence of any one of these conditions, the act would not constitute killing, even though death should follow, and will not entail any evil effect on the perpetrator.

moral guidelines to do what one should do as a moral being, and their social and religious consequences. It is up to each individual to keep them according to his capacity and the situation in which he finds himself. Therefore one should not be surprised to see that the precept against killing and other basic precepts are violated by many Buddhists based on their irrelevance to modern life.[20] Yet at the same time there are also those who use the precepts to form an all important motive for moral life and as a means to accumulate merit to ensure good rebirth. The belief in an afterlife makes the difference in dealing with life problems including the issue of right or wrong, good or bad.

The new development of non-ES cell research to find stem cells from other sources such as umbilical cord blood and bone narrow is an important effort to balance the need for the relief of suffering and moral obligation of non-maleficence. Time, effort, and money should be spent more on adult stem cell research making it possible to produce different types of cells, such as blood cells, muscle cells and neural cells. Many scientists in different counties are engaging in this research and some have reached the stage of conducting clinical experiments. However, adult stem cell research still has a long way to go. It is not an easy task. But it is worth doing it for the sake of science and humanity. It is a way to fulfill our obligations as moral beings.

Conclusion

Human genetic research has great potentials. It can change the direction of human history. It can remodel human beings according to the desire of what humans want to be. What we have discussed so far is only in terms of possibilities whether positive or negative. But we cannot be sure how things will turn out and what will be best for humanity and the planet itself. This creates hope for some and fear and anxiety for others. It is our hope that the progress of human genetic research and regenerative medicine will help us to be healthier and live longer. The fear arises mostly from the possibilities of the abuse of science to satisfy human desires without thinking of the dangers that it can bring about. The massacre of millions by the Nazi's eugenic movement still haunt human conscience. This kind of tragedy happened because scientists themselves are not saints, nor can they be so. They are people of mixed motivations, a combination of strengths and weaknesses, like all of us. Though their motivation in research may be altruistic at the beginning, it may be replaced later by selfishness and greed, e.g. to gain fame and fortune, personal prestige, higher paying position, public recognition and more research grants. We may cite the notorious case of Prof. Hwang Woo-suk of Seoul National University in Korea who, for personal gains, claimed falsely that he had succeeded in producing ES cells from a cloned embryo.

[20] The five basic precepts (Pañca-sīla) are: (1) abstain from killing; (2) abstain from stealing; (3) abstain from sexual misconduct; (4) abstain from lying, and (5) abstain from intoxicants.

In this time of hopes and fears, science should be incorporated with the cherished values of compassion and human rights which will enable it to be really beneficial to humanity. Science and ethics are not at opposite ends. They can be together and supplement each other. The urgent need to humanity is to have scientists with conscience and compassion. These two moral qualities can set the moral limits to what one will do to others in the present and future generations. With these limits scientists engaged in genetic research, for example, will not go on beyond therapeutic cloning to produce a new human being, even though they have the capacity to do so, due to the unsafe procedures and possible negative effects such cloning will be to humanity.

Compassion is as important as an inquisitive mind in sciences. It involves both beneficence and non-maleficence. With compassion as motivation, science and technology will be steered towards the optimum relief of human suffering, and the welfare of humanity in general. This needed compassion and conscience are already inherent in our hearts and minds, but are in a slumber-like state because of our egoism. They need to be awakened, nurtured and developed. Engagement in bioethics is one important means to make scientists aware of the moral implication of their research and their social responsibility. Religious commitment is another way to strengthen conscience, enabling them to resist temptations, do what is right and good, and seek what is truly of value. The Buddhist way is the practice of mindfulness meditation. For more than 2,500 years this Buddhist meditation has been used for moral, development and spiritual transformation. The practice enables us to be fully aware of our thoughts *(sati)*, destructive or positive, and know *(sampachañña)* what we are doing whether it causes happiness or suffering to others, and our reason for doing it. Moral rules and principles can regulate human behavior, but they cannot change the human heart. Meditation is specifically designed to transform the heart by purifying it from all taints of egoism and make it a loving and compassionate heart. It will then be alert to the suffering of others and spontaneously ready to relieve that suffering. With this compassionate heart one becomes a *bodhisattva* who lives and works not for himself but for the welfare of others.[21]

On the spur of the moment the question may be raised about the relevance of meditation to the work and life of the scientist. As already discussed meditation can make a difference in the life of the practitioner. The scientist is usually aware of what he is doing. But meditation can expand this awareness to include the sense of interdependence between self and others, and humanity as a whole. Consequently meditation can enable the scientist to curb his ego desires and develop compassion. The mindful scientist will then not only be concerned with his research but also

[21] *Bodhisattva* refers to the one who is a Buddha-to-be. It is believed that prior to becoming a Buddha, Kotama the Buddha was bodhisattva in 550 successive lives practicing all kinds of Perfections, e.g. Perfection of Wisdom, Perfection of Compassion. The stories of the Buddha's previous lives are collected in *Jatakas Stories*.

with its application and impacts. When scientific knowledge and compassion are woven together, the possibilities of modern science and technology, leading man to destruction, could be minimized and even eliminated. There is a bright future for humanity.

References

1 Wallace AB: The Taboo of Subjectivity. London, Oxford University Press, 2000, pp 75–97.

2 Davids R (transl): The Book of the Kindred Sayings. London, Luzac, 1954, vol III, p 157.

3 Sadakata A: Buddhist Cosmology: Philosophy and Origins. Tokyo, Kōsei, 2004.

4 Keown D: The Nature of Buddhist Ethics. London, Macmillan, 1999.

5 Davids R (transl): The Book of the Kindred Sayings. London, Luzac, 1952, vol II, p 23.

6 Ratanakul P: Human cloning: Thai Buddhist perspectives; in Roetz H (ed): Cross-Cultural Issues in Bioethics: Example of Human Cloning. Amsterdam, Rodopi, 2000, pp 203–214.

7 Ratanakul P: To save or let go: Thai Buddhist perspectives; in Keown D (ed): Contemporary Buddhist Ethics. London, Curzon Press, 2000, pp 169–182.

Pinit Ratanakul
College of Religious Studies, Mahidol University, Salaya
Puthamoltoll 4
Nakornpathom, 73170 (Thailand)
Tel. +66 2 800 2630/39, Fax +66 2 800 2659, E-Mail pinitratanakul2@hotmail.com

Author Index

Brahier, G. 1

Childress, J.F. 28

Engelhardt, Jr., H.T. 40

Gehring, P. 60
Graf, F.W. 52
Green, R.M. 118

Lindpaintner, K. 1

Mieth, D. 82
Müller, H. 74

Nor, S.N.M. 129

Peters, T. 108
Pfleiderer, G. 1

Ratanakul, P. 138
Rehmann-Sutter, C. 12

Schockenhoff, E. 101

Subject Index

Abortion, sex differences in ethics 61, 62
Agrippa 41, 42
Anthropological reductionism 56–58

Barth, Karl 115
Bateson, Gregory 21
Bio-idolatry, avoidance 114, 115
Biopolitics, church involvement 70–72
Biopower
  embryo ethical status 95–97
  freedom of procreation 97–99
  overview 82–86
  politics of language 86–91
  theological arguments 91–95
Blair, Tony 15
Bleich, David 125
Blueprint, genetic 15, 112
Brain death, bioethicization of church positions 67
Broyd, Michael J. 126
Buddhism
  compassion 140, 141, 149
  embryo self-determination rights 144, 147
  ethics goals 140
  genetic manipulation and human dignity 144, 145
  human action and freedom 141, 142
  Human Genome Project views 143, 144
  humanistic ethics 142
  overview of GenEthics 9, 10, 138
  reproductive cloning views 145, 146
  science analogies 139
  stem cell research views 147, 148
  suffering and cessation 139, 140

Cahill, Lisa Sowle 116
Calvinism 114, 115
Catholicism
  Christian view diversity 45, 46
  human life as biological entity 62, 63
  overview of bioethics 6, 7
  power of churches 61, 62
  response to bioethics conferences and declarations 94, 95
Childress, James 2, 3, 28
Chimerism 112–114
Christianity, *see also* Catholicism, Orthodox Christianity, Protestantism
  bioethicization of church positions 65–68
  plurality of religious views 45–49
Clinton, Bill 15, 16
Clinton Commission 87, 89
Creation 54, 55

Darwinism 18, 74, 114
Deane-Drummond, Celia 111, 112
Declaration of Independence 54
Dignity, *see* Human dignity
Dolly, sheep cloning 82–85
Dor Yeshorim program 118

Embryo research
  Buddhism and self-determination rights 144, 147
  ethical status of early embryo 95–97
  Islam views 129, 132, 133
  Judaism views 120–124
  potentiality argument 25
Emden, Jacob 124
ENCODE project 77
Engelhardt, H. Tristram 3, 4, 40
Essentialism 108, 109

Feminism 63, 64, 97, 98
Frank, Adam 116

Gehring, Petra 4, 5, 60
Gene, definition 17
Gene myth
  bio-idolatry avoidance 114, 115
  chimerism 112–114
  origins of thought 108, 109
  playing God 110
  support 110–112
Genetic chance, rights 106, 107
Genetic screening, right not to know 102–104
Genetic text, authoring 22–24
Genome sacredness, *see* Gene myth
Graf, Friedrich Wilhelm 4, 52
Green, Ronald M. 8, 118

Healthcare directives, church views 68, 69
Hemophilia A, heredity 77
Hermeneutics, genetics 14, 15, 26
Human dignity
  Buddhism 144, 145
  conceptions
    standard of duties 29–31
    standard of virtue 31–33
  genetic enhancement and inequalities 33–35
  growth of importance 36
  overview 28, 29
  prospective technology assessment 35, 36
  public precautionary process 35–38
  reproductive cloning 30–32, 83, 84
  wisdom of repugnance 32
Human Genome Project
  Buddhism views 143, 144
  overview of ethics 15, 16, 109, 111
Human Rights Convention on Biomedicine 88, 89, 94
Humanae Vitae 62, 63

In vitro fertilization
  Islam views 131, 132, 134
  Judaism views 118, 121
  overview of ethics 83, 97, 98
Interdisciplinary dialogue, fostering 79, 80
Islam
  embryo views 129, 132, 133
  in vitro fertilization views 131, 132, 134
  maqasid al-Shariah 129–131, 135, 136
  overview of bioethics 9, 129, 130
  reproductive cloning views 131, 134
  sources for ethical views 130, 131
  therapeutic cloning views 135

Jacob-Monod model 17–19
Jacobovits, Immanuel 123
Jonas, Hans 84, 107
Judaism
  development and nascent life views 120–124
  in vitro fertilization views 118, 121
  life-preserving principles 119, 120
  overview of GenEthics 8, 9, 118, 119
  prenatal diagnosis views 118, 119, 121, 124, 126
  sources for teachings 119
  technology openness 124–126

Kaneko, Kunihiko 19
Kant, Immanuel 44, 45, 65, 83, 89
Kass, Leon 2, 31, 32, 34
Kay, Lily 24

Lac operon 17, 18

Macklin, Ruth 28–30
Maqasid al-Shariah 129–131, 135, 136
Maslahah mursalah 131
Mendel, Gregor 74
Mieth, Dietmar 6, 82
Morality
  consensus illusion 43, 44
  diversity 41, 42
  plurality of religious views 44–49
Müller, Hansjakob 5, 74

Natural birth, rights 104–106
Naturalism 115, 116
Neumann-Held, Eva 20
Nor, Siti Nurani Mohamed 9, 129
Novak, David 121

Organ transplantation, bioethicization of church positions 66–68
Orthodox Christianity
  Christian view diversity 45, 46
  overview of bioethics 46–49
Oyama, Susan 20, 21

Peters, Ted 7, 8, 108
Pikuach Nefesh 119, 120, 122, 126

Pinker, Steven 33
Playing God
chimerism 112–114
definition 110
Polysemy 24
Pregnancy crisis counseling 94
Prenatal diagnosis
Judaism views 118, 119, 121, 124, 126
overview of ethics 83, 87, 97–99, 102, 103
President's Council on Bioethics 2, 29, 31, 34, 110
Procreation, freedom 97, 98
Protestantism
Christian view diversity 45, 46
human life as biological entity 64, 65
overview of bioethics 7, 8
Public precautionary process, human dignity 35–38

Ratanakul, Pinit 9, 138
Rehmann-Sutter, Christoph 2, 12
Religious language
abuse in ethics 54–56
interpretation 53, 54
Reproductive cloning
human dignity concerns 30–32, 83, 84
Islam views 131, 134
politics of language 86–91
RNA, variants 20
Rosner, Fred 118, 126

Samsara 141, 142
Schockenhoff, Eberhard 7, 101
Secular bioethics 40, 41
St. Basil 47, 48
Stem cell research, Buddhism views 147, 148
Systems approach, genomics 2, 21, 22, 25

Tay-Sachs disease 118, 124
Theologians, role in bioethics discussions 75, 76
Therapeutic cloning
Islam views 135
overview of ethics 87

Ummah wasatah 135
Unterman, I.J. 123

van Wolde, Ellen 92
Vitalism 18

Waldenberg, Eliezer 124
White, Lynn, Jr. 35, 36
Woo-suk, Hwang 148
Worthing, Mark 114